Testimonials

"*Pro Life, Pro Woman, Pro Science* addresses the overarching issues in the debate on abortion that divides our country today. As a clinical OBGYN who has been on both sides of this issue, practicing life affirming medicine for over 30 years, my hope is that it encourages readers to move those 18 inches beyond the rhetoric of intellectual division to the real source of connection and transformation that is found in the center of our chests, the human heart. Well done Bonita Pratt"

- **Dr. John Bruchalski FACOG, Former Abortionist, Author, and Founder of Tepeyac OBGYN**

"Pro-life, Pro-woman, Pro-science offers a grammar for understanding the legal, moral, and medical implications of abortion. Readers who are either unfamiliar with the pro-life position or who want to better present this position in dialogue with others will learn much from this work. I know that I did."

- **Professor Timothy O'Malley, Author and Theologian**

"Bonita Pratt excellently explains through this work what is needed today: for pro-life Americans to be equipped with the truth in a society that too easily embraces falsehood as fact. While the right to life is not a nuanced topic — the science that points to conception at fertilization is copious and clear — humans are complex beings with an instinct to think deeply about matters of morality. If you are seeking a way to think through your beliefs about human life and how to defend them, the right resource has reached your hands."

- **Prudence de Bernardo,**
Former Anchor of EWTN Pro-life Weekly

"This work is an absolute must-read for anyone interested in engaging others in meaningful dialogue around this critical topic. It is a speakers manual, providing a level of professional research collated in a user-friendly way that is unavailable anywhere else. In preparing for presentations through the years, I've spent hours trying to find the information so beautifully organized and presented in this work. The sections on chemical abortion are particularly important in this post-Roe era where internet marketers are selling death that requires only an internet connection to purchase. And, despite working over 30 years as an OB/GYN, the work centered on the breast cancer/abortion link is better than anything I've seen. I can't recommend this book strongly enough."

- **Dr. Christopher Stroud OB/GYN CFCMC,**
Founder of Fertility and Midwifery Care Center
and Holy Family Birth Center

Pro-Life, Pro-Woman, Pro-Science

An Integrative Guidebook

Bonita Pratt

En Route Books and Media, LLC

Saint Louis, MO

⊛ENROUTE
Make the time

En Route Books and Media, LLC

5705 Rhodes Avenue

St. Louis, MO 63109

Cover credit: Michael Murphy

Copyright © 2025 Bonita Pratt

ISBN-13: 979-8-88870-420-2

Library of Congress Control Number:

Available online at https:catalog.loc.gov

To my wonderful husband, Tony

Table of Contents

1.

Introduction

Growing up, being pro-life was easy. You drop off a box of diapers at a pregnancy center, hold a sign with a catchy saying once a year, and talk with friends about favorite baby names. As beautiful as that world is, we all get older and begin to hear more complicated arguments, passionate emotions, and countless sources all claiming to be authoritarian.

I will never forget sitting in my college classes and watching ardently pro-choice professors vocalize their opinions about abortion as though they were facts. I still remember looking around the room at all my classmates as no one dared to stand up to the professors, but also no one seemed to be readily agreeing with them either. As I started talking to my classmates individually, they confided that while they did not agree with the professor, they also did not feel well equipped to respond. It was then that I started to realize how many well intentioned people of good character want to defend the sanctity of unborn life but feel intimidated into silence by the strong statements and quick stats that the pro-choice side will rattle off.

Post college, working in multiple facets of the pro-life field, I saw an even greater need for an easy go-to handbook for people who do not have the time to do all the research themselves. So, this is that book. It is written to guide you along pro-life / pro-choice conversations, so that you can feel confident and well-prepared to reach out to others. God has given each of us such beautiful and different gifts. For those of you who use your talents to give glory to God through engineering, medical care, childcare, sales, construction, marketing, entertainment, government, and everything else, thank you for your daily sacrifices. And, thank you for letting me use my talents of research, organizing, and writing to help grow the pro-life movement! Together, we will create a lovely, well-rounded culture of life!

2.

Summary of Supreme Court Cases

Understanding the history of abortion in the United States, particularly the game changing Supreme Court rulings, is necessary for contextualizing the issue. It increases our understanding of how we got to where we are now and what key factors, words, and ideas forever changed the trajectory of the debate.

Roe v. Wade:

On behalf of Jane Roe (Norma McCorvey), attorneys Sarah Weddington and Linda Coffee brought forth a court case against Wade County and the Texas criminal abortion laws which made abortion illegal, except in cases of life of the mother. Weddington and Coffee advocated for unrestricted access to abortion for all women. The majority opinion ruled the case in their favor (7 - 2), based on the Ninth and Fourteenth Amendments, on January 22, 1973. Thereby, abortion became legalized in all the states (*Roe* Syllabus 113).

Doe v. Bolton:

Doe v. Bolton was decided on the same day as *Roe v. Wade* and broadened the access to abortion that *Roe* had just given. Specifically, it ruled that abortions did not have to be approved by a hospital abortion committee, nor performed at an JCAH accredited hospital. Further, there was no longer a requirement for two independent physicians to confirm that the abortion was needed, nor for the patient to be a resident of the state wherein the abortion took place, in this case, Georgia (*Doe* 201). These restrictions were deemed to be "invalidating as an infringement of privacy and personal liberty" (*Doe* Syllabus 179). Thus, *Roe* made abortion legal, and *Doe* made it widely available.

Planned Parenthood v. Casey:

On June 29th, 1992, the court reaffirmed *Roe v. Wade* with its ruling on *Planned Parenthood v. Casey* (5 - 4). The opinion of the court focused on *stare decisis* and a woman's right to liberty as the basis for a right to abortion. One key difference is that *Planned Parenthood v. Casey* dismissed *Roe's* trimester framework and instead focused on viability as the defining factor. It declared that a State could not place an "undue burden" on a woman's access to abortion before viability, but that the State could enact restrictions after viability (*Planned Parenthood* Syllabus 837).

Specifically, it ruled that a married woman no longer needed to notify her husband before obtaining an abortion, but that a state could still require informed consent of the woman before the procedure, a 24 hour waiting period, parental consent for minors with the option of a judicial bypass, and specific reporting standards for abortion clinics. This ruling left much confusion and disagreement about how to practically carry this out at the state level and what exactly constituted an "undue burden" (*Planned Parenthood* Syllabus 837).

Stenberg v. Carhart

"Undue burden" (*Stenberg* Syllabus 916) came into the spotlight yet again, when Dr. Leroy Carhart challenged Nebraska's law that prohibited D&X, commonly known as partial birth, abortions. Dr. Carhart claimed that such a ban placed an "undue burden" upon him and his patients (*Stenberg* Syllabus 916, 938 & 946). On June 28, 2000, the court narrowly ruled in favor of Dr. Carhart, 5 - 4. In response the federal branch took action and Congress enacted the Partial Birth Abortion Ban Act of 2003, prohibiting this type of abortion from continuing to take place anywhere (United States 2003).

Dobbs v. Jackson Women's Health Organization:

On June 24, 2022, the majority opinion ruled in favor of the Mississippi Department of Health (5 - 3 with 1 concurring in judgment), declaring that the state's ban on abortions after 15 weeks of gestation (pre-viability) was valid. The ban did have an exception for "a medical emergency or in the case of a severe fetal abnormality" (*Dobbs* Opinion 6). This contradicted the old standing that states could only restrict abortion access after the point of viability. In their ruling, the majority reversed the decision in *Roe v. Wade* and returned the power of decision making back to the individual states.

3.

Comparison of Key Elements between *Dobbs v. Jackson* and *Roe v. Wade*

While all these cases played a pivotal role in the abortion issue, two in particular stand out as needing a deeper analysis: *Roe v. Wade* and *Dobbs v. Jackson Women's Health Organization*. These two cases, that broadly legalized abortion and then reversed that legalization, go into great detail about the historical context, Constitutional text, women's rights, fetal development and when life begins, the proper role of the Court, and implications for the future. Each of these considerations warrants its own analysis.

History:

Before *Roe* gave an overarching ruling legalizing abortion, states were able to individualize their abortion laws. Since the Constitution was enacted, citizens were able to elect their state officials to create laws that reflected the majority opinion. (*Dobbs* Opinion 1). *Roe* ended this individuality by demanding that all states must legalize abortion, regardless of the views of its citizens and elected officials. This included the 30 states that currently had laws making abortion illegal at any point in pregnancy (*Dobbs* Opinion 2). Looking back even earlier,

"At common law, abortion was criminal in at least some stages of pregnancy and was regarded as unlawful and could have very serious consequences at all stages. American law followed the common law until a wave of statutory restrictions in the 1800s expanded criminal liability for abortions. By the time of the adoption of the Fourteenth Amendment, three-quarters of the States had made abortion a crime at any stage of pregnancy, and the remaining States would soon follow" (*Dobbs* Opinion 16).

Roe then nationally legalized abortion, despite these states' and their citizens' concerns. These concerns, however, did not go away. More than half of the states have directly petitioned the Supreme Court to reverse the erroneous decision made by *Roe* and allow them to restrict abortion as they and their citizens see fit (*Dobbs* Opinion 4). Considering the above, the *Dobbs* majority decided that "[t]he inescapable conclusion is that a right to abortion is not deeply rooted in the Nation's history and traditions" (*Dobbs* Opinion 25).

Stare Decisis:

The *Dobb's* majority's opinion also spent a considerable amount of time looking into the proper application of *stare decisis* - the Latin phrase for "to stand by things decided" - and the idea of respecting past decisions and precedents. They point out that *Roe*, itself, was not accurately based on any past precedent "because all of the precedents *Roe* cited... were critically different for a reason that we have explained: None of those cases involved the destruction of what *Roe* called "potential life"" (*Dobbs* Opinion 37). This new life is of the utmost importance and completely changes the core of the case.

Regarding respecting *Roe* as settled precedent, the *Dobb's* majority asserts, "*stare decisis* is "not an inexorable command,… and it 'is at its weakest when we interpret the Constitution'" (*Dobbs* Opinion 39). *Roe* undeniably relied on an interpretation of the Constitution, not the specific text. The *Dobbs* majority points out that the Constitution does not directly dictate abortion, or even privacy, as a right (*Dobbs* Opinion 9). They also offer the reminder that the court should not be swayed by public pressure and opinion, even if part of the public is very vocal about supporting the previous Court's decision (*Dobbs* Opinion 67). Further, many other critical and influential Supreme Court rulings reversed previous rulings (*Dobbs* Opinion 40). In his concurring opinion, Justice Kavanaugh highlights the fact that, "Every current Member of this Court has voted to overrule precedent. And over the last 100 years… every one of the 48 Justices appointed to this Court has voted to overrule precedent" (*Dobbs* Kavanaugh 6). Thereby, Justice Kavanaugh points out the hypocrisy of the dissenting judges who insisted that *Roe* must be reaffirmed based on precedent. Considering these three things, the majority concludes that simply because *Roe* and *Casey* were ruled in one way, does not necessitate that they remain the law of the land forever.

- "The Court has no authority to decree that an erroneous precedent is *permanently* exempt from evaluation under traditional *stare decisis* principles" (*Dobbs* Opinion 68).
- "*Stare decisis*, the doctrine on which *Casey's* controlling opinion was based, does not compel unending adherence to *Roe's* abuse of judicial authority. *Roe* was egregiously wrong from the start. Its reasoning was exceptionally

weak, and the decision has had damaging consequences. And far from bringing about a national settlement of the abortion issue, *Roe* and *Casey* have inflamed debate and deepened division" (*Dobbs* Opinion 5-6).

Constitutional Text:

In *Roe v. Wade,* the court based their legal standing on "the concept of personal "liberty" embodied in the Fourteenth Amendment's Due Process Clause; [and] in personal, marital, familial, and sexual privacy said to be protected by the Bill of Rights" (*Roe* Opinion 129). They concluded that a woman was at liberty to abort a pregnancy if she so desired and that such an action would be private and, thereby, not limited by the state. In doing so, they chose to assume that abortion fell under the categories of liberty and privacy, even though, "the Constitution makes no reference to abortion, and no such right is implicitly protected by any constitutional provision" (*Dobbs* Opinion 5). Surprisingly, the dissenting judges agreed with this stating, "The majority says (and with this much we agree) …In 1868, there was no nationwide right to end a pregnancy, and no thought that the Fourteenth Amendment provided one" (*Dobbs* Dissent 13). Yet, the Fourteenth Amendment has continually been used to promote the pro-abortion agenda.

Privacy:

In his dissent from *Roe,* Justice Rehnquist pointed out that, "A transaction resulting in an operation such as this is not "private" in the ordinary usage of that word" (*Roe* Rehnquist 172). This is due to the fact that an abortion does not just involve a woman's body, but also an unborn human being and member of society. This changing of terminology from *Roe's* "potential life" to *Dobb's* "unborn human being" rhetoric highlights the shift in perspective of how the unborn are viewed and, now, respected. (*Roe* Opinion 150) (*Dobbs* Opinion 5).

Liberty:

This understanding, of personhood in the womb, necessarily affects the way "liberty" is also interpreted. Early in our nation's history, President Lincoln pointed out that, "we all declare for Liberty; but in using the same word we do not all mean the same thing" (*Dobbs* Opinion13). Playing off of this duality of terminology, the *Dobb's* judges wisely noted the limitations of one's liberty in that, "while individuals are certainly free *to think* and *to say* what they wish about 'existence,' 'meaning,' 'the universe,' and 'the mystery of human life,' they are not always free *to act* in accordance with those thoughts" (*Dobbs* Opinion 30-31). In this way, people may choose to think about the unborn in any manner they choose, but, like everyone else, their actions have to be aligned with the law. If people were to have unrestricted liberty, then they could use the Fourteenth Amendment to justify all sorts of heinous actions, such as murder or rape. In order to have a functioning society, it is necessary to have "ordered liberty (which) sets limits and defines the boundary between competing interests" (*Dobbs* Opinion 31). Since liberty can be interpreted in a vast number of ways and since the Constitution does not specifically address abortion, the *Dobb's* judges advise against using judicial power to make a final decision whether or not abortion is an American right (*Dobbs* Opinion 14). Instead, they return the power back to the people and the individual states. In his concurring opinion, Justice Kavanaugh states that, "The Constitution is neutral and leaves the issue for the people and their elected representatives to resolve through the democratic process" (*Dobbs* Kavanaugh 2). In his dissent from *Roe*, Justice Rehnquist goes a step further and asserts that in legalizing abortion, "the Court necessarily has had to find within the scope of the Fourteenth Amendment a right that was apparently completely unknown to the drafters of the Amendment," since at the time of the amendment "three-quarters of the States had made abortion a crime at any stage of pregnancy" (*Roe* Rehnquist 174) (*Dobbs* Opinion 16). Clearly, the creators of the amendment did not intend to directly counteract the current laws at the time. Consequently, the *Dobb's* majority concluded that the supposed right to abortion was not inherently protected in the Constitutional rights of privacy and liberty, and that individual states should have the freedom to choose to restrict and prevent abortions in accordance with respecting the unborn human life.

Beginning of Life:

The defining factor of the entire abortion debate is a difference in beliefs about when a human person's life begins. If the fetus were not a human person, then abortion would be no different from other medical procedures. There would be no death involved and no immorality associated with the action. However, the fetus being a small human person, means that the termination of that life is murder and that the government has a duty to make such practices illegal.

Shockingly, the majority opinion in *Roe* decided that, "We need not resolve the difficult question of when life begins" (*Roe* Opinion 159). Thereby, they dismissed and discredited the very cornerstone of the entire issue. Simultaneously, they tried to create a distinction between a human being, biologically, and a human being with personhood, whose rights deserve to be protected. They cited postnatal uses of the term "person" in the Constitution and claimed that, "conception is a "process" over time, rather than an event" (*Roe* Opinion 157, 161). However, they admit that, "If this suggestion of personhood is established, the appellant's case, of course, collapses, for the fetus' right to life would then be guaranteed specifically by the Amendment" (*Roe* Opinion 156-157). In essence, they would agree that human personhood is an overarching factor on the issue, but yet, they refuse to look into the scientific explanation for the beginning of human life and personhood. Thereby, they leave the most fundamental question in the shadows so as to avoid coming across the research that would discredit their entire ruling.

Fetal Development:

Because of its vast implications and importance, the judges of *Dobbs'* majority opinion revisit the issue of fetal development as it relates to personhood. They argue against the subjective and ever-changing definition of personhood that *Roe* asserted and that the dissenting judges currently support. "According to the dissent, the Constitution *requires* the States to regard a fetus as lacking even the most basic human right—to live—at least until an arbitrary point in a pregnancy has passed. Nothing in the Constitution or in our Nation's legal traditions authorizes the Court to adopt that 'theory of life'" (*Dobbs* Opinion 38-39). Instead, the *Dobbs* opinion looks at the modern science for fetal development and points out the difference in scientific understanding and knowledge from what was previously known (*Dobbs* Opinion 7). Before modern scientific advances in technology, it was extremely difficult, or impossible, to have knowledge about a pregnancy before quickening (maternal ability to feel fetal movements) (*Dobbs* Opinion 21). Therefore, it was difficult to determine whether or not there was another "person" inside the woman. Once quickening occurred and the fetus's movement could be felt, then it became undeniably evident that it was a human person inside the womb. However, quickening does not happen at a universally specific time. Rather, "quickening is a maternal sensitivity and depends on the mother's own fat, the position of the placenta and the size and strength of the unborn child" (Motion). Now, even with the advances in technology showing early pregnancy detection and prenatal development, pro-choice advocates chose to not advance their definition for when life begins and personhood is established, because doing so would discredit their claim for a right to abortion.

The *Dobbs* dissent does exactly this and tries to reframe the narrative in terms of bodily autonomy for the women. "Every-one, including women, owns their own bodies. So the Court has restricted the power of government to interfere with a person's medical decisions or compel her to undergo medical procedures or treatments" (*Dobbs* Dissent 21). This statement completely ignores the fetus's right to "own their own body" and utterly dismisses the idea that the fetus is also a human person with rights (21). The dissent aligns itself with the original *Roe* ruling and specifically disagrees with the majority's understanding that conception and fertilization happen at the same time. "Today, the Court discards that balance. It says that from the very moment of fertilization, a woman has no rights to speak of" (*Dobbs* Dissent 2). In this way, they begin the scare tactics and terrifying wording that they then use throughout the rest of their dissent, in order to drum up the fear factor in the American people.

Fear:

Fear is one of the most powerful tools that exists and can be masterfully manipulated in order to shape a society's beliefs and actions. Both the Dissent in *Dobbs* and the Opinion of the Court in *Roe*, capitalize on this tactic of fear mongering in order to paint the issue in unrealistic and drastic extremes to the American people, particularly American women. The *Dobbs* Dissent is littered with false and grossly exaggerated claims, along with words chosen precisely to play off of many people's current fears.

- "A State will be able to impose its moral choice on a woman and coerce her to give birth" (*Dobbs* Dissent 3).
- "One result of today's decision is certain: the curtailment of women's rights, and of their status as free and equal citizens" (*Dobbs* Dissent 4).
- "The consti-tutional regime we enter today erases the woman's interest and recognizes only the State's" (*Dobbs* Dissent 12).
- "It forces her to carry out the State's will, whatever the circumstances and what-ever the harm it will wreak on her" (*Dobbs* Dissent 30).

- "An American woman is 14 times more likely to die by carrying a pregnancy to term than by having an abortion" (*Dobbs* Dissent 22).
- "Burdens of pregnancy, childbirth, and parenting" (*Dobbs* Dissent 43)
- "They may lose not just their freedom, but their lives" (*Dobbs* Dissent 51).
- "After today, young women will come of age with fewer rights than their mothers and grandmothers had" (*Dobbs* Dissent 54-55).

These statements portray pregnancy, childbirth and parenting as a sort of slavery and death sentence for the woman. They pit the mother against her child, claiming that either the child must be killed from abortion or the mother's happy life will necessarily be killed instead.

This intimidation theme is nothing new and is merely a continuation of the approach used by the judges 49 years prior in *Roe*. One particular paragraph stands out:

"The detriment that the State would impose upon the pregnant woman by denying this choice (abortion) altogether is apparent. Specific and direct harm medically diagnosable even in early pregnancy may be involved. Maternity, or additional off-spring, may force upon the woman a distressful life and future. Psychological harm may be imminent. Mental and physical health may be taxed by child care. There is also the distress, for all concerned, associated with the unwanted child, and there is the problem of bringing a child into a family already unable, psychologically and otherwise, to care for it. In other cases, as in this one, the additional difficulties and continuing stigma of unwed motherhood may be involved" (*Roe* Opinion 153).

While this statement is painting a very bleak outlook for pregnant women, it has very little standing. Most of the adverse effects listed are hypothetical and "may" or may not happen. There is also an appalling lack of sources for all of the harm and negative physical, emotional and social consequences that they insist will occur. In his Dissenting Opinion from *Roe*, Justice White puts the issue in a broader perspective, pointing out, "At the heart of the controversy in these cases are those recurring pregnancies that pose no danger whatsoever to the life or health of the mother but are, nevertheless, unwanted for any one or more of a variety of reasons — convenience, family planning, economics, dislike of children, the embarrassment of illegitimacy, etc" (*Roe* White 132). While some tragic cases and unfortunate scenarios certainly do exist, we must not allow the pro-choice side to exaggerate their frequency and severity in order to make the objectively evil wrong of abortion considered a legally protected right in every single scenario.

Women's Rights:

Coupled with their fear tactics, the pro-choice judges in both *Roe* and the *Dobb's Dissent* warp feminist ideologies in order to portray motherhood in an enslaving light and abortion as a woman's only chance at equality. In their dissent, Justices Breyer, Sotomayor and Kagan make the far-reaching claim that abortion is the method by which women can choose when and if to raise children and that, "Without the ability to decide whether and when to have children, women could not—in the way men took for granted—determine how they would live their lives, and how they would contribute to the society around them" (*Dobbs* Dissent 24). Essentially, they are arguing that women will have terrible and inferior lives, unless they protect their "right" to kill their children. In the original *Roe* ruling, the judges conceded that "appellant and some amici argue that the woman's right is absolute and that she is entitled to terminate her pregnancy at whatever time, in whatever way, and for whatever reason she alone chooses. With this we do not agree" (*Roe* Opinion153). Yet, on the same day, their ruling in *Doe v. Bolton* essentially gave the green light to abortion for whatever reason (*Doe* Opinion 201).

It is this mentality of abortion on demand that persists in pro-choice America, today. Because of this mentality and wide-spread perception that abortion rights are women's rights, the judges in *Dobb's* had to be very clear that their reversal of *Roe* did not directly harm women and infringe upon their due rights. They point out that, "the 'goal of preventing abortion' does not constitute 'invidiously discriminatory animus' against women" and that the "laws regulating or prohibiting abortion... are governed by the same standard of review as other health and safety measures" (*Dobbs* Opinion 11). They also discuss the multitude of ways by which society has changed to be supportive of pregnant women; including maternity leave, pregnancy Medicaid, "safe haven" laws, well established adoptive homes, a greater understanding of fetal development and laws against pregnancy discrimination in the workforce (*Dobbs* Opinion 33-34). They also point out women's current political influence. "Our decision returns the issue of abortion to those legislative bodies, and it allows women on both sides of the abortion issue to seek to affect the legislative process by influencing public opinion, lobbying legislators, voting, and running for office. Women are not without electoral or political power" (*Dobbs* Opinion 65).

Ironically, the abortion conglomerate was only able to secure a woman's supposed "right to abortion," by usurping the freedom of two women: Norma McCorvey (Jane Roe) and Sandra Bensing Cano (Mary Doe). Neither women ended up aborting their pregnancies and neither were fully involved in the court process. Joseph Dellapenna wrote extensively how the lawyers "Weddington and Coffee kept McCorvey out of the loop; McCorvey had virtually no involvement in the litigation other than providing an almost fictional plaintiff" (Dellapenna 679). Cano's experience was eerily similar. She later revealed in an affidavit that, "I neither wanted nor sought an abortion. I was nothing but a symbol in Doe v. Bolton with my experience and circumstances discounted and misrepresented" (Affidavit 35). During this time, when she was pressured into aborting, she fled to Oklahoma (Affidavit 39). She had been misled into believing that her lawyer, Margie Pitts Hames, was helping her with divorce papers and custody issues (Affidavit. 35-36). She even goes so far as to assert that, "I am certain the signature on the affidavit that said I wanted an abortion was not mine. I never saw that affidavit until the records were unsealed" (Affidavit 39). The manipulation and coercion that these two women endured is truly appalling and the polar opposite of upholding women's rights and dignity.

Irrelevant additions:

Since only the loosest of interpretations can claim to find a right to abortion in the actual Constitutional text, the judges of *Roe* turned to outside sources for justification. They reference ancient European laws - even though such cultures supported infanticide - the English statutory law, the position of the American Medical Association, the American Public Health Association, and the American Bar Association, among others (*Roe* Opinion 130, 137, 141, 144, 146) (*Dobbs* Opinion 47). They specifically targeted sources to promote the pro-choice agenda. Notably, *Roe* included the American Public Health Association's recommendations regarding abortion: "Rapid and simple abortion referral must be readily available...counseling should be to simplify and expedite the provision of abortion services...sympathetic volunteers... may qualify as abortion counselors...(and) Contraception and/or sterilization should be discussed with each abortion patient" (*Roe* Opinion 144-145). Clearly, these provisions are meant to make abortion access as easy and quick as possible, instead of giving women time to process their thoughts and information from medical professionals about other options. At the same time, the judges also discredit the longstanding, pro-life Hippocratic oath by claiming that it is irrelevant and highly controversial (*Roe* Opinion 130-132). All considered, the question must be asked: why are all these positions given so much weight and a total of 16 pages, when none of them have to do with the wording of the Constitution's text? The judges in *Dobbs* point this out, observing that, "After cataloging a wealth of other information having no bearing on the meaning of the Constitution, the opinion concluded with a numbered set of rules much like those that might be found in a statute enacted by a legislature." (*Dobbs* Opinion 2). Truly, the *Roe* court acted in a manner due to the legislative branch, not the judicial.

Five Factors:

In order to properly justify their overruling of *Roe, Doe, and Casey*, the majority in *Dobbs v. Jackson* focused on five main factors: "the nature of their error, the quality of their reasoning, the 'workability' of the rules they imposed on the country, their disruptive effect on other areas of the law, and the absence of concrete reliance" (*Dobbs* Opinion 43).

1. Nature of the Error: The majority begins by asserting that *Roe* was not only wrong, but "egregiously wrong and deeply damaging," along with being "outside the bounds of any reasonable interpretation" (*Dobbs* Opinion 44). Thereby, they emphasize the severity of the case. They continue, "*Roe* was on a collision course with the Constitution from the day it was decided, *Casey* perpetuated its errors, and... wielding nothing but 'raw judicial power,' ... the Court usurped the power to address a question of profound moral and social importance that the Constitution unequivocally leaves for the people" (*Dobbs* Opinion 44). In this way, the multifaceted nature of abortion is brought back into light, contradicting *Roe's* claim that, "the abortion decision in all its aspects is inherently, and primarily, a medical decision" (*Roe* Opinion 166).

2. Quality of Reasoning: The majority gave an exhaustive list of
    the plethora of ways by which *Roe's* reasoning was faulty.

> "It failed to ground its decision in text, history, or
> precedent. It relied on an erroneous historical narra-
> tive; it devoted great attention to and presumably re-
> lied on matters that have no bearing on the meaning
> of the Constitution; it disregarded the fundamental
> difference between the precedents on which it relied
> and the question before the Court; it concocted an
> elaborate set of rules, with different restrictions for
> each trimester of pregnancy, but it did not explain
> how this veritable code could be teased out of any-
> thing in the Constitution, the history of abortion
> laws, prior precedent, or any other cited source; and
> its most important rule (that States cannot protect fe-
> tal life prior to "viability") was never raised by any
> party and has never been plausibly explained" (*Dobbs*
> Opinion 45-46).

Additionally, the majority points out how *Casey* was poorly
based on *stare decisis*, failed to strengthen *Roe's* reasoning
and left much confusion about what exactly "viability" and
an "undue burden" meant (*Dobbs* Opinion 46, 56). The majority found many problems with the "viability" distinction;
including that viability has changed with time, is different
based on the location of the woman and the hospital's technology, can only be estimated and not known for certain,

and has "nothing to do with the characteristics of a fetus" (*Dobbs* Opinion 51-53).

3. Workability:

Workability refers to "whether it can be understood and applied in a consistent and predictable manner" (*Dobbs* Opinion 56). The "undue burden" test presented by *Casey* has been extremely difficult to implement on a day to day level, due to a great number of varying interpretations of its meaning (*Dobbs* Opinion 56). In order to provide clarity, the *Dobbs* majority specifically noted that, "legitimate interests include respect for and preservation of prenatal life at all stages of development, ... the protection of maternal health and safety; the elimination of particularly gruesome or barbaric medical procedures; the preservation of the integrity of the medical profession; the mitigation of fetal pain; and the prevention of discrimination on the basis of race, sex, or disability" (*Dobbs* Opinion 78). Every state now has the explicit authority to enact laws according to the above criteria, even if abortion access decreases as a result.

4. Disruptive Effect:

Roe and *Casey* have had a detrimental ripple effect on multiple "legal doctrines" (*Dobbs* Opinion 62). They have added confusion and misguided understanding of several important factors and cases. Particularly, "They have ignored the Court's third-party standing doctrine... disregarded standard *res judicata* principles... (and) distorted First Amendment doctrines" (*Dobbs* Opinion 63). This is quite the opposite from the clarity and uniformity that *Roe* and *Casey* promised to provide.

5. Absence of Concrete Reliance:

   Finally, the court looked at whether or not the reversal of *Roe* would cause serious practical complications due to people planning their lives around the current laws. They concluded that, "getting an abortion is generally 'unplanned activity,' and 'reproductive planning could take virtually immediate account of any sudden restoration of state authority to ban abortions'" (*Dobbs* Opinion 64). In other words, even if abortion were to become illegal, women would be able to quickly choose methods for family planning.

After considering all of the above, the majority concluded that, "procuring an abortion is not a fundamental constitutional right because such a right has no basis in the Constitution's text or in our Nation's history" (*Dobbs* Opinion 77).

Implications:

Lastly, the majority in *Dobbs*, emphasized multiple times that their ruling was only applicable to reversing *Roe's* ruling on abortion and that the *Dobbs* ruling should not be "misunderstood or mischaracterized" to be permission to reverse other previous court rulings

(*Dobbs* Opinion 66). This was due to the dissent's insistence that a reversal of *Roe* would signal in a reversal of the legalization of inter-racial marriage, contraception, the morning after pill, same-sex marriage, IVF and miscarriage treatment (*Dobbs* Dissent 19, 20, 24, 36-37). Clearly, these constitute a wide range of issues. While some of these do not involve unborn human life, some of them do. Consequently, the dissent rightfully acknowledges that, "'[R]easonable people… could also oppose contraception; and indeed, they could believe that 'some forms of contraception' simi-larly implicate a concern with 'potential life' (*Dobbs* Dissent 24). This is in reference to the morning after pill and other similar forms of contraception that act as an early abortifacient; ending the life of the child before implantation. Considering these factors, Justice Thomas alone, in his concurring opinion, called for the court to "reconsider all of this Court's substantive due process precedents, including *Griswold* (contraception), *Lawrence* (same-sex intercourse) and *Obergefell* (same-sex marriage) (*Dobbs* Thomas 3).

4.

Value of Life

Before going into the nitty gritty of abortion procedures, situations, and complex arguments, a basic question needs to be answered: "Does life have value? And why?" If life does not have value, then abortion suddenly becomes not a big issue at all. Pro-choicers will oftentimes distinguish between the human DNA of the fetus and the human DNA plus personhood of children and adults. This distinction played a pivotal role in *Roe v. Wade* as Justice Blackmun acknowledged that, "If this suggestion of (fetal) personhood is established, the appellant's (Roe's) case, of course collapses, for the fetus' right to life would then be guaranteed specifically by the Amendment" (*Roe* Opinion 156-157). So when and what is it that gives one personhood and, thereby, value and a right to life?

The pro-lifer would quickly respond without hesitation that personhood is established at the moment of conception. On the other hand, the pro-choice side has a whole spectrum of answers ranging from vague, ill-defined points in pregnancy up to birth. Some of the ways by which they determine personhood is by consciousness, experiences, memories, sense of community and ability to create. Let's briefly consider each of these points. Consciousness: adults experience periods of unconsciousness all the time from sleep, fainting, anesthesia, temporarily being knocked out from a sporting event, or coma. Further, different adults have different levels of comprehension and self-recognition, but clearly they are not different levels of persons. Experiences: The baby in utero has many experiences, even if they are somewhat different from the types of experiences adults have. Their experiences include receiving nutrients, kicking and moving limbs, responding to voices and music, and many others. This also begs the question, if experiences make one a person, then what if an adult has very limited experiences or only bad experiences. Are they less of a person? Obviously not. Memories: While the preborn memory is evidently not as developed as an adult's long term memory, it does not mean that they have no memories. The baby hears and remembers the mother's voice so much in utero that (s)he will kick and move in response and will also respond specifically to the mom's voice at birth. Also, if memories determine personhood, are the elderly or other folks with short term memory loss, less of a person? Absolutely not. Interaction / Community: The baby very much interacts with his/her mother in utero. As they progress in pregnancy, they interact more and more with the father and doctors as well, as they make sounds and put pressure on the womb. Additionally, many pregnancies are with multiples. These babies are interacting with each other in utero from the very beginning. Ability to create: What is it that we need to create in order to be considered human persons? Each person has different gifts and can use their talents to create different things. Even an in utero baby can create a smile or feelings of love. While all of these markers are terribly inconsistent, they are also impossible to precisely, scientifically determine. We do not have the medical knowledge to say a precise time mid-pregnancy or birth that each of these traits begins. Thus, one's respect for the value of unborn personhood is entirely based on the unknown. Additionally, the embryo, as a human being, has the natural capacity for each one of these criteria from the very beginning of his/her existence, even if that capacity has yet to be activated.

This distinction between a human body and later personhood is utterly absurd. While living on earth, you cannot be separated from your body. Your body is a part of you and you are composed of your body (and soul). One cannot reasonably claim that the living embryo body is not a person, any more so than one could claim that (s)he is not the person culpable for his/her adult body's actions. You cannot separate your body from yourself. Your personhood is linked with your body.

What makes human personhood so unique is that we are an inseparable combination of body, intellect and an immortal soul. Animals have a body and a sort of animated soul, but it is not an immortal soul, nor do they have the capacity for reason and logic like we do. Every human person has intrinsic dignity ipso facto that they are a person. This "intrinsic dignity does not rely on the choices of anyone, but is something that remains whether or not someone is recognized to have it" (Kaczor 769). It is an inherent quality that no one can take away. Sometimes it can be confused with attributed or inflorescent dignity, which considers the accomplishments of the person and whether they are thriving in life (Kaczor 770). While these types of dignity are somewhat subjective, intrinsic dignity is innate and present in each and every person. It can never be taken away. Further, theologically speaking, we are made in the image and likeness of God, our Creator. This is yet another pivotal distinction between human persons and any other living thing. While Christianity teaches us about our inherent and undeniable dignity and worth; an overarching respect of human life, and thereby a right to life, has long transcended many religions, cultures, and times. There is an overarching Natural Law that instills in our hearts the recognition that life is valuable. Faith, science and reason are all in agreement. "Though faith is above reason, there can never be any real discrepancy between faith and reason. Since the same God who reveals mysteries and infuses faith has bestowed the light of reason on the human mind, God cannot deny himself, nor can truth contradict truth" (Catholic Church 159). The understanding of the value of human life and the dignity of the person from the very beginning of his/her existence is sublimely rooted in both faith and reason. This enables us as the pro-life generation to have deep and meaningful conversations with those who want to debate abortion from any perspective: scientific, religious, philosophical or any other!

5.

Why Life begins at Conception

When convincing someone that life begins at conception, it is typically easiest to persuade them by showing them why life could not logically begin at any other marker, such as birth or viability. Ask them when they believe life begins and start with the responses for that particular stage and then slowly work your way back further and further, earlier into the pregnancy until conception/fertilization is the only option left. At that point, then explain why not only is it the only option left, but also why it is such a logical conclusion and standard.

Not Birth:

Pro-choicers sometimes assert that a baby is not a person until birth because that is the point at which it is completely independent of the mother. This proposition places extreme importance on a location change of a few inches and a time change of a couple minutes. A baby, who is 40 weeks in the womb, has the same basic level of development in that moment as in one hour from then - after being born. The difference of only a couple minutes of time, changes whether or not some people consider this baby a human person or just a disposable fetus. Nothing has intrinsically changed about the baby. There is just a slight external change in location. However, even the change in location is small, equaling only a few inches. Why is location the determining factor for whether or not someone has the right to life? We, as a society, do not go around killing other people simply because we do not like the place where they live. The baby has the same level of development regardless of whether or not (s)he has traveled down the birth canal. Further, babies are born at a wide range of gestational ages. While 40 weeks is the typical marker for being full term, many babies have been born in the second trimester and many are born even at 42 weeks. For example, if two babies were conceived at the same time and one was born at 27 weeks, while the other was born at 41 weeks, the second would be at risk of abortion for an additional 14 weeks. The first baby would obviously not be allowed to be "aborted" after birth up until the original due date. Ironically, it is the baby who stayed in the natural, normal, and healthy environment for the appropriate amount of time, who is the one at risk of dying.

Frequently, people who believe that life begins at birth, do so because that is when the baby takes his/her first breath. Since stopping breathing means that someone has died, people then assert that the fetus is not a person until (s)he can breathe. However, it is faulty reasoning to assume that the same marker for death is the same marker for being alive. We must ask ourselves: why is it important to breathe? Obviously, it is to receive the oxygen that our body needs to sustain itself and function. While the unborn baby is not breathing through her/his nose or mouth, that baby is receiving oxygen through the umbilical cord and placenta, as is perfectly natural and normal for that environment and stage of development. It is through this different, but completely healthy, mode of receiving oxygen that the baby's body is able to sustain itself and function appropriately. "Transition to extrauterine life is characterized by changes in circulatory pathways, initiation of ventilation and oxygenation via the lungs instead of the placenta" (Morton). It is because of this oxygen transfer through the umbilical cord that babies can be born via waterbirth without drowning. A baby does not take his/her first breath until after (s)he has been exposed to air. Therefore, a baby can be born underwater and be perfectly okay, even without breathing air for the first little bit (How to Water Birth) (Tan 373, 379). This leads to the question: if breath determines life, then can you "abort" a baby after delivery but before the first breath? Absolutely not!

If your conversation partner remains adamant that abortion is permissible all the way up till birth, then it can be beneficial to explain what exactly a late term abortion entails. In a late term abortion, called an induction abortion, labor and delivery are induced. However, instead of delivering an alive baby, the child dies before delivery. See section below for details on induction abortion. Therefore, if the woman has to go through labor and delivery anyway, why not deliver an alive baby? Further, delivering a baby normally can be healthier for the mother, since her body will naturally be ready to deliver and the end result is not being forced. Emergency c-sections are another option for a woman who needs an immediate delivery. If the woman is concerned about raising a child and not just childbirth, this could be a good opportunity to segue into the option of adoption. See details below for how to discuss adoption.

Not Viability:

Viability is the point at which the fetus/baby can survive outside of the womb. The pro-choice argument goes like: so long as the fetus is dependent upon the mother for survival, it cannot be considered a person with rights; but if the fetus could survive on its own, outside of the womb, then the fetus is its own person. The first thing to remember is that there is nothing wrong with a pre-viability fetus. A fetus or embryo in the early stages of development acts in a completely normal and natural way. The fetus/baby is doing exactly what (s)he is supposed to do for that specific time in life. The fetus/baby is able to interact with the environment, grow, and develop perfectly on schedule. The fetus/baby is perfectly viable so long as (s)he is kept in the natural environment of the womb. Viability outside the womb is simply another stage in the progression of development. Even babies, who are viable outside of the womb and do not need extra medical help, are still dependent on their parents for feeding and basic care. They would die if left alone. Likewise, some adults even need extra medical care, such as an inhaler or feeding tube, in order to stay alive; though clearly they have reached the level of viability. Even some perfectly healthy people need aid because of their environment, such as astronauts. Astronauts are extremely healthy and fit people, but need extra equipment to stay alive while they are in space. Similarly, premature babies need help functioning outside of their normal, natural environment - the womb. The majority in *Dobbs* accurately points out that, "The most obvious problem with any such argument is that viability is heavily dependent on factors that have nothing to do with the characteristics of a fetus" (*Dobbs* Opinion 51). These outside factors are technology, time, location, and subjective interpretations of viability.

Technology:

40 weeks of pregnancy is the standard length for determining a baby's due date and declaring that the pregnancy has reached full

term and the baby will be fully viable. However, millions of babies are routinely born prematurely (before 37 weeks) and survive (Preterm). These premature babies are given extra medical care in order to keep them alive and healthy. As medical technologies, scientific understanding and medicines advance, doctors are able to help babies at earlier and earlier gestational ages. Therefore, the unborn baby's ability to be viable outside the womb is dependent on current medical technologies. Consequently, saying that a baby is not a person with rights before viability is attained, is essentially saying that one person's technological skills determine whether or not someone else has a legally protected right to life.

Time / Always Changing:

Since technological advances in the medical field are continuously occurring, the stage of viability keeps getting pushed back earlier and earlier. It is not a consistent standard, but rather one that has evolved with time and continues to do so. According to this logic, a 27-week fetus would not be considered a person with a right to life a few hundred years ago, but would be considered one now. Or, a 19-week (before current external viability) fetus could be aborted now, but could be considered to be a viable baby in the future. The majority in *Dobbs* concurs that viability is dependent on the "state of neonatal care at a particular point in time. Due to the development of new equipment and improved practices, the viability line has changed over the years" (*Dobbs* Opinion of the Court 52). Thus, viability is again found to be an inconsistent standard for determining personhood and a right to life.

Location:

Just as advances in medical technology have been made over time, there are also differences in technology based on location. Hospitals in third world countries have vastly fewer resources than hospitals with NICUs in the United States. Therefore, a baby who is viable in the United States, might not be viable in Africa; even though the child has the same level of development (Preterm). The World Health Organization reports, "More than 90% of extremely preterm babies (less than 28 weeks) born in low-income countries die within the first few days of life, yet less than 10% of extremely preterm babies die in high-income settings" (Preterm). A similar argument could be made for an expansive city hospital vs. a small rural one. Is it logical that an unborn child in a small town with fewer medical resources, would not have a right to life, but if that same child were born in a big city with an expansive health network, then (s)he would? The *Dobbs* majority asks a similar question, "On what ground could the constitutional status of a fetus depend on the pregnant woman's location" (*Dobbs* Opinion 52)?

Rates and Subjectivity:

Viability is difficult to determine because it occurs on a spectrum. Babies born prematurely have different levels of success at continuing to live. In their informed consent brochure, the Indiana State Department of Health gave the following percentages for chance of survival based on gestational age: 34wk >98%, 30wk >95%, 27 wk >90%, 26w 80% - 90%, 25w 50% - 80%, 24wk 40% - 70%, 23w

10% - 35%, 22w 0% - 10%, 21w or less 0% chance of survival (Abortion Consent 6, 2022). Other research from the Journal of the American Medical Association is much more hopeful and gives a 28% survival rate, "with active treatment," at 22 weeks and a 55% survival rate at 23 weeks (Richter). Even with this knowledge, a couple of issues arise.

First, each fetus/baby's health and circumstances are unique. A range differing 30% (see weeks 24 and 25) highlights how little about viability is known and how it can be disagreed upon drastically. The majority in *Dobbs* highlights this saying, "It is thus "only with difficulty" that a physician can estimate the "probability" of a particular fetus's survival" (*Dobbs* Opinion 53). Further, "settling on a "probabilit[y] of survival" that should count as "viability" is another matter... Is a fetus viable with a 10 percent chance of survival? 25 percent? 50 percent" (*Dobbs* Opinion 53)? Therefore, the definition of viability could change based on the conclusion that one wants the definition to support.

Secondly, it is very important to fact check any source that just does not quite sound right. While the Indiana State Department of Health is a reputable source, it sounds odd that the brochure implies that a 21 week old fetus/baby has an absolute 0% chance of survival. Further research shows that, though rare, a few babies have survived after being born this early. So, while the number might be very close to 0%, it is not an absolute 0%. In 2024, a little boy, Nash Keen, was born at 21 weeks 0 days. He was delivered to the University of Iowa Health Care and received top level medical care for 6 months before being released to go home. He holds the Guinness World Record for most premature baby to live (Millward). The previous record was held by Curtis Zy-Keith Means, born at 21 weeks 1 day, and Richard

Huchinson, born at 21 weeks and 2 days; both in the summer of 2020 (Millward). A month earlier, Richard Huchinson had just set the premie record at 21 weeks and 2 days (Millward). Further back in 2014, a baby girl, Lyla, was born at 21w 4d. She survived and is a happy, healthy toddler now. Sadly, Dr. Ahmad, who saved Lyla's life, was rather pessimistic in his interview with *Today*. *Today* reported, "people need to be very cautious about concluding from a single case that routinely resuscitating babies born in the 21st week of gestation is the best approach, Ahmad warned. Don't assume one positive outcome will be the outcome for other infants, he noted" (Pawlowski). He went on to shed a shocking new light on the subject. "At this time, resuscitating infants who are born in the 21st week of gestation is not standard practice anywhere in the world. Even for those infants born in the 22nd week of gestation, there continues to be significant disagreement regarding the best course of action due to the high mortality and substantial risks for long term disability" (Pawlowski). This begs the question, why in the world is it not standard practice to resuscitate infants born at 21 weeks? Simply because they are at an increased risk of sickness, disability or even death, does not mean that you should withhold life-saving care. The worst possible outcome, death, will certainly become reality if no alternative is offered. Thus, the 0% survival rate becomes a self-fulfilling prophecy. Along the same line, a 2022 Stanford Medicine article shared new "analysis shows infants even at the lowest gestational ages — 22 and 23 weeks — might live if they are actively resuscitated… 'There has been a shift toward considering a more active initial treatment in prenatal discussions with families over the past several years in light of increasing data to support this approach,' [Dr. Hintz] said" (Richter). Again, there is this assumption that standard protocol is not to resuscitate and have an "active initial treatment" plan (Richter). Such a standard is abhorrent and a true stain on the medical industry. Dr. Hintz then talks about how new data shows the potential for these little babies to continue living. How did they get this data? It could only have been the result of brave parents fighting for the lives of their tiny babies and choosing active resuscitation and treatment from the earliest days and then proving that it actually can work! Again, if no one is trying to help these tiny babies, of course they are going to die and then be prematurely labeled as non-viable!

Not Ability to Feel Pain:

The Susan B. Anthony List recently conducted a poll which showed that "55% of likely voters say they are more likely to support a 15-week limit on abortion when they learn that an unborn child has the capacity to feel pain" (Where do Americans stand). While the baby is being killed, regardless of whether or not (s)he can feel pain during the process, the knowledge that pain is being inflicted upsets many people. Pain is a very common human experience, so the fetus's ability to feel pain demonstrates his/her humanity. Once humanity is established, it is extremely difficult to propose arguments that justify abortion. Since the ability to feel pain is such a humanizing factor, how pain is defined is critically important. People can twist the definition of pain in order for it to align with their argument. The American College of Obstetricians Gynecologists give the following definition, "The experience of pain 'requires conscious recognition of a noxious stimulus.'" (Gestational Pain). Not surprisingly, they are vocal supporters of abortion and want "to integrate abortion as a component of mainstream medical care, and to oppose and overturn efforts restricting access to abortion" (Abortion Policy). The British Medical Journal proposed a different definition of pain as something that "cannot be explicitly expressed or measured…We propose that the fetus experiences a pain that just is and it is because it is, there is no further comprehension of the experience, only an immediate apprehension" (Derbyshire and Bockmann). This definition offers a more holistic understanding of pain. Similarly, Srinivasa Raja M.D., a professor of Anesthesiology and Neurology and Director of Pain Research at the Johns Hopkins University School of Medicine points out that, "Inability to communicate does not negate the possibility that a human or nonhuman animal experiences pain" (Raja et al.). Contrasting these definitions shows how people can come to vastly different conclusions and, thereby, frame the argument in completely different contexts. This leads to a wide range of opinions on when a fetus/baby first experiences pains from 7 weeks gestation all the way to birth:

- "Pain experience…occurs outside the womb" (Derbyshire)
- "A human fetus does not have the capacity to experience pain until after at least 24–25 weeks." (Gestational Pain)
- "By 20 weeks' gestation, an unborn child has the physical structures necessary to experience pain… (and) evade(s) certain stimuli in a manner" (HB 481).
- "The mechanisms for physiological endocrine reactions to pain are certainly in place (at 18 weeks and more at 20 weeks.)" (Myers et al.)
- "Hormonal stress response at 18 weeks." (Fetal Awareness 7)
- "Fetal surgeons and anesthesiologists, routinely administer fetal analgesia at increasingly earlier gestations in the second

trimester (>14 weeks gestation) to ameliorate pain and improve outcome" (Thill)
- "Evidence, points towards an immediate and unreflective
pain experience mediated by the developing function of the
nervous system from as early as 12 weeks." (Derbyshire and
Bockmann)
- "Specialized nerve terminals, nociceptors, are likely to detect
surgical tissue damage from… 10 weeks." (Fetal Awareness
5)
- "Nerve fibers grow into the fetal spinal cord from 8 weeks."
(Fetal Awareness 5)
- "Noxious stimuli are first sensed by peripheral nociceptors
in the perioral area at 7.5 weeks gestation" (Thill)
- "Projections from the spinal cord reach the brainstem and
thalamus beginning at 7 weeks gestation" (Thill)
- "An intact spinothalamic projection might be viewed as the
minimal necessary anatomical architecture to support pain
processing… at seven weeks' gestation." (Derbyshire)

Clearly, the first assertion, that claims that pain can only be experienced after birth, is specifically set up to be a talking point in
support of abortion. Why do they choose to ignore the interaction
of the fetus with the environment of the womb? The many different
markers of pain development highlight how complex the issue of
pain is and how limited our current medical understanding is. However, if your friend says that (s)he does not support abortions once
the fetus/baby can feel pain, then you could ask her/him, "Since fetuses can potentially feel pain as early as 7 weeks gestation, do you
agree that abortions after that point should be illegal?" This could be

a great opportunity to find some common ground before continuing on with the conversation.

While the inability of the fetus/baby to feel pain might potentially lessen one's feelings of guilt, it does not change the fact the baby will die at the end of the abortion. It cannot change the reality or morality of the situation. The ability to feel pain is simply another stage of development. What makes this bodily function more important than any other? If the other person remains insistent that pain is the crucial difference, you could try asking, "If pain is your marker, are you okay with an abortion at 9 months so long as the fetus/baby was given a pain killer first?" If they respond no, ask them why that would not be okay. One can also bring up that some individuals have a medical condition that prevents them from physically feeling pain (Daneshjou et al.). These individuals are no less human than those of us who can feel pain. Overall, look for common ground that you can build off of. If they respond yes, then it is no longer a fetal pain question and you have to circle back up to previous talking points on the beginning of life.

Not Heartbeat:

Astoundingly, those who work in the abortion industry frequently acknowledge the embryo's beating heart. They even discuss harvesting the fetal heart tissue after an abortion. Perrin Larton of Advanced Bioscience Resources Inc., a company that facilitates the harvesting of fetal tissue, admitted under oath that sometimes, "I can see hearts that are not in an intact p.o.c. (embryo/fetus) that are beating independently" (Planned Parenthood Staff) (Overview). This is an utter travesty!

While the abortionists will readily discuss post-heartbeat abortion, this makes many people uncomfortable since one very common pro-choice definition of personhood is the heartbeat. Since the heart beats through the entirety of a person's life outside of the womb and stops beating when someone dies, the pro-choicers conclude that the heartbeat is a necessary function in determining life. However, they fail to consider that the definition of death does not have to be the same as, but reverse of, the definition of life. How could the embryo be not alive (aka dead), if time alone can bring it into life with no extraordinary help? The heart beginning to beat is simply another natural stage in early fetal development. Since the heart beginning to beat is such a sticking point for so many people, pro-choicers have attempted to claim that the heartbeat does not start until much later in the second trimester. Thus, women who did not want to abort after a heartbeat still have a plethora of time to go through with the abortion. The American College of Obstetrics and Gynecology states, "Until the chambers of the heart have been developed and can be detected via ultrasound , it is not accurate to characterize the embryo's or fetus's cardiac development as a heartbeat" (ACOG Guide 2025). Their 2023 definition included the timeframe of "roughly 17-20 weeks of gestation" (ACOG Guide 2023). However, a 2019 article published by the Oxford Cardiovascular Clinical Research Facility points out that there is the "formation of the 4-chambered heart by gestational week 7 (from conception)" (Tan 374). However, even before the 4 - chambers are developed, the heart has begun to beat. The same Oxford article discusses the timeline; "The initiation of the first heartbeat via the primitive heart tube begins at gestational day 22, followed by active fetal blood circulation by the end of week 4 (from conception)" (Tan

373). Specifically, "During week 4 (from conception), the heart tube undergoes rightward looping, with its posterior region moving anteriorly" (Tan 375). Many other sources confirm this finding (Morton) (HB 481) (Donovan Weeks 6-8). Even though the heart beats at such an early point in the pregnancy, the embryo still meets all the basic criteria of being a living human being before then: homeostasis, organization, metabolism, responsiveness, movement, reproduction, growth, differentiation, respiration, digestion, and excretion (Body Functions).

Homeostasis:

Homeostasis is "the ability of an organism to maintain constant internal conditions" (Bartee). This is necessary in order for the human being to stay alive (Body Functions). "Homeostasis depends on the body's ceaselessly carrying on many activities. Its major activities or functions are responding to changes in the body's environment, exchanging materials between the environment and cells, metabolizing foods, and integrating all of the body's diverse activities" (Body Functions). The yolk sac aids the embryo's homeostasis by enabling the embryo to receive gasses and nutrients (Yolk Sac). It is this "continuous supply of fetal nutrients rather than hormones (that) control fuel homeostasis" (Rao). Later on, when the placenta takes over, "[it] plays a critical role in the maintenance of homeostasis in pregnancy" (Ishida et al.). Even before implantation, the blastocyst partakes in homeostasis. "In preparation for implantation... the blastocyst also improves homeostatic regulatory mechanisms" (Vergaro). There is also homeostasis within individual tissues and even within individual cells from the very beginning of human life (Homeostasis).

Organization / Comprised of Cells:

The blastocyst is "composed of hundreds of cells" at the time of implantation and has only continued to quickly grow in the following weeks (Cell Division). Each cell has internal organization between its nucleus, cell membrane and cytoplasm (NCI Dictionary). Further, each type of cell is organized throughout the embryo. By the 4th week from conception, there are multiple types of cells organized together and working to continue development. There are already present brain cells, nerve cells, eye cells, ear cells, and limb cells (Stages First Trimester).

Metabolism:

"Metabolism is a broad term that includes all the chemical reactions that occur in the body", specifically the use of energy (Body Function). The embryo's metabolism is evidenced by its growth, development and functioning abilities. It is so important that, "metabolism is considered a key determinant of embryo competence and viability" (Vergaro). Even before the embryo stage is reached, metabolism plays a key role. "Changes (regarding the blastocyst) are energy dependent, and therefore underpinned by specific metabolic pathways" (Vergaro). "The newly formed child has a remarkable degree of metabolic autonomy" (Motion).

Responsiveness:

Responsiveness means reacting with one's environment (Body Functions). A very early example of this occurs when the blastocyst implants in the mother's uterus. The blastocyst is responsive to the environment of the uterus in a way that differs from its response to being in the mother's fallopian tube. Even in cases of ectopic pregnancy, where the blastocyst implants somewhere other than the uterus, the blastocyst is still responding to and interacting with its environment, just in an unhealthy and fatal way. Implantation is just one example of this responsiveness. In general, "the embryo is a dynamic structure that can be affected by the interaction with the surrounding environment" (Vergaro).

Movement:

The blastocyst traveling through the fallopian tube and to the uterus is one very clear example of early movement. The small embryo then makes small movements by the yolk sac and, later, the placenta. There is also movement of individual cells within the embryo. On an even smaller scale, there are also molecular movements within each cell (Body Functions).

Reproduction:

Reproduction can occur in humans in two ways, sexually or at the cellular level. Obviously, unborn babies cannot sexually reproduce, but neither can born infants or young children. The human body has to continue developing and go through the stage of puberty

in order to be able to sexually reproduce. However, cells are repro-
ducing from the beginning of a pregnancy. At the moment of conception, the new human being is a single celled organism. "After 30
hours or so, it divides from one cell into two. Some 15 hours later,
the two cells divide to become four. And at the end of 3 days, the
fertilized egg has... 16 cells." By the time of implantation, the blastocyst is now "composed of hundreds of cells" (Cell Division). The
cells continue to multiply and the embryo becomes the length of
1/6th of an inch by the fourth week after conception, which is when
the heartbeat can first be detected (Stages First Trimester).
Growth:

At 6 weeks gestation (4 from conception) the embryo has already
grown drastically. At conception, the embryo was a single cell. At 4
weeks gestation (2 from conception), the embryo is approximately
1/100th of an inch. By 6 weeks gestation: "The embryo is about ⅙ -
in long and has developed a head and a trunk. Structures that will
become arms and legs, called limb buds, begin to appear. The brain
develops into five areas and some cranial nerves are visible. The eyes
and ear(s) begin to form. Tissue forms that develop into the vertebra
and some other bones" (Stages First Trimester).

Differentiation:

Differentiation is a "process by which unspecialized cells change
into specialized cells with distinctive structural and functional characteristics" (Body Functions). As mentioned above, by the time the
heart beats; brain cells, nerve cells, eye cells, ear cells, and limb cells

are all present and have specific, different functions (Stages First Trimester). A few specific examples of this specialization are:

- Two weeks after conception heart cells are beginning to gather together (Tant 375).
- "Between 2 and 3 weeks' gestation (from conception), the yolk sac initiates fetal [red blood cell production)]" (Morton).
- At four weeks from conception: "Rudimentary blood moves through the main vessels" (Stages First Trimester).

Respiration:

While most people think of breathing when they think of respiration, scientifically, "respiration refers to all the processes involved in the exchange of oxygen and carbon dioxide between the cells and the external environment" (Body Functions). At the very beginning of pregnancy, the blastocyst / embryo receives oxygen through passive diffusion. At this stage in development gas exchange is aided by the yolk sac, specifically the chorion (Morton) (Donovan Yolk Sac). Further, "the blastocyst is one of the most active tissues... characterized by exponential growth, expansion of a blastocoel, increased oxygen use and oxidative metabolism" (Gardner & Phil). "By the end of gestational week 3 (from conception), passive oxygen diffusion becomes insufficient to support metabolism of the developing embryo, and thus the fetal heart becomes vital for oxygen and nutrient distribution" (Tan 373). Respiration also occurs on a cellular level in all living cells (Cellular Respiration). Therefore, respiration occurs immediately after conception when the baby is a single celled organism.

Digestion:

Even before implantation, the embryo has nutritional needs that are being met. "The relative availability of nutrients within the female reproductive tract mirrors precisely the changing requirements of the preimplantation embryo at each stage of development" (Gardner & Phil). Also early on, "oxygen and nutrients… can be delivered to the embryo or fetus by the maternal circulation" (Delgado et al. 4). Further, the yolk sac is crucial in the embryo's ability to receive nourishment and discard waste. Right at the beginning of implantation, the chorion begins to develop (Genbacev et al.). "The chorion also develops from the yolk sac and functions to nourish the developing embryo" (Donovan Yolk Sac). As such, the embryo is able to receive and digest nutrients, even though the process looks different from adult digestion.

Excretion:

"The yolk sac also gives rise to the allantois at three weeks gestation (from conception). The allantois is… involved in removing nitrogenous waste and products and is associated with the development of the urinary bladder" (Donovan Yolk Sac)

Some sources also consider adaptation a separate characteristic. Unsurprisingly, the embryo also meets this criteria before the heart has started to beat.

Adaptation:

Adaptation refers to the fact that, "all living organisms exhibit a 'fit' to their environment" (Bartee). The reason why the embryo can survive inside the womb, but not outside of the womb this early on, is because its development is directly related with its environment. "To facilitate survival in the hypoxemic intrauterine environment, the fetus possesses structural, physiological, and functional cardiovascular adaptations that are fundamentally different from the neonate" (Tan 374).

Not Implantation:

Lastly, the pro-choice side tries to avoid recognizing a pregnancy, let alone the personhood that is causing the pregnancy, until implantation. Implantation refers to when the blastocyst implants within the uterine lining. It occurs approximately 6 days after fertilization, though the exact timing is nearly impossible to pinpoint (Conception). Alabama and California are two examples of state codes defining pregnancy differently. Alabama's code defines pregnancy as beginning with fertilization, whereas California's health and safety code defines pregnancy as beginning with implantation (2012 AL) (2009 CA). The American College of Obstetrics and Gynecology agrees with California that, "intrauterine pregnancy begins when a fertilized egg implants itself in the uterus" (ACOG Guide). Note that they are very careful to specify "intrauterine pregnancy." Yes, the intrauterine phase of the pregnancy begins when the blastocyst implants within the uterus, but the pregnancy is already in existence before then.

This distinction about when a pregnancy begins is important because it relates to birth control usage and what pills are deemed to be actual abortion pills. The American College of Obstetrics and Gynecology asserts that "emergency contraception," which blocks the blastocyst from implanting, "is not an abortifacient; it does not end a pregnancy" (ACOG Guide). This is because they choose to ignore the pregnancy up until the point of implantation. Remember, at the time of implantation the embryo has already grown to be 1/100 of an inch (Stages First Trimester). While they can make this claim as long as they want, it is simply a lie to say that forms of contraception, such as Plan B, do not act as early abortifacients (Plan B). However, contraceptives that block implantation are currently not legally labeled as abortifacients, allowing them to be differentiated from abortions and to remain legal regardless of what restrictions states pass on early abortions. Since HCG surges only after implantation, a woman is not able to test positive for pregnancy until after implantation. Therefore, she cannot know definitively that she is pregnant until after implantation has occurred. Since abortions only refer to post-implantation pills and procedures, a woman cannot have an abortion until after implantation. Thus, if someone believes that abortion is only okay before implantation, the conversation would switch from one about legalized abortion to one about abortifacient birth control.

Nonetheless, there is still plenty of evidence showing why a human person is present even before implantation. Once conception / fertilization has occurred, a unique human organism, with its own DNA, has already come into existence. While the cells are rapidly multiplying in the days between conception and implantation, location is the major change from right before implantation and right after implantation. The baby, in the zygote / blastocyst stage, is traveling through the fallopian tube to the uterus where (s)he can implant in the uterine lining. Like we discussed in the birth and viability sections; why would one's location of a few inches be the deciding factor in life versus death? Like the other stages, implantation is simply another natural stage in fetal development.

An important note about this stage in development, is that it is nearly impossible to precisely pinpoint. While a woman's body may help hint at a pregnancy's timeline, there is not any current medical technology that is able to definitively determine whether or not implantation has occurred. Consequently, if one is okay with pre-implantation abortifacients, but is not okay with post-implantation abortions, than their respect for life is a bit arbitrary and clouded since there is a period of time in which a woman and her doctor are not sure if implantation has occurred or not.

While implantation is important because it gives us knowledge of the embryo, it does not intrinsically change the nature of that child or automatically instill value into that tiny person's life. One's knowledge of another person should not determine whether or not the other tiny person is actually a human person with dignity and rights.

Yes, Fertilization/Conception:

Considering all of the above, the only remaining point for a new human being, with full personhood, to come into existence would be fertilization/conception. It is the only consistent way to determine when life begins because all of the above forms of measure-

ment are subjective and many are changing. No other way of deciding the beginning of life can be pinpointed to a specific moment and
event that dramatically changes things. The moment, when the
sperm fertilizes the egg and conception occurs, is when it is no
longer just the woman's body. Her egg could never just spontane-
ously turn into a zygote. At that moment of conception, the tiny
baby, still in the zygote stage, already has his/her own unique set of
DNA. That set of DNA is a unique and beautiful blend of the mother
and the father, not a mere replica of either. The zygote's DNA already will determine the eye color, hair color, sex, and future height
of the baby.

While the simplicity of truth in this answer speaks volumes,
some pro-choicers try to have a "gotcha" moment by bringing up
identical twins, arguing that their existence means that not every human being is unique or that not all life starts at conception. First off,
looking at the technical medical definition is helpful. These twins are
not scientifically labeled as identical, but rather as "monozygotic"
because in the early stage of pregnancy there is one zygote (Bruder
et al. 763). In the days following fertilization, the zygote splits into
two - now there are twins (Kyono et al.). Though monozygotic twins
start off from the same zygote, they quickly develop differences.
McGill University recently published an article stating, "The DNA
of monozygotic twins tends not to be 100% identical" (Identical
twins). The differences between the twins are due to genetic, epigenetic and environmental factors (Bruder et al. 763) (Identical twins).
Genetically, "the enzyme tasked with copying our DNA makes mistakes, and if these mistakes go uncorrected, they stick around as mutations" (Identical Twins). While not all of these "mutations" are

bad, they do account for differences between twins. Epigenetics refers to "ways in which our bodies have evolved to dictate when a particular stretch of our DNA should be active, and while these epigenetic markers are inherited from our parents, they can be reversed and change over time, and that means that monozygotic twins can drift away from one another because their epigenetics change independently of one another" (Identical Twins). Environmentally, twins have different experiences that not only shape personal preferences, but also their health history and exposure or immunity to different diseases. All of these factors highlight the uniqueness of the individual. No one is an exact copy of another person, not even an "identical" twin. We are all made gloriously unique and have a special gift to share with the world.

There are a couple different ways of thought considering when the lives are formed. Option 1 is the belief that there is one person alive at the moment of fertilization and continuing on through pregnancy and that the second person comes into existence at the time of the split and then also continues to live. Option 2 is that just like conjoined twins share much of the same body and organs, so too would the persons share the same zygote body at the beginning and then naturally separate later (Horn). Twins are not delivered at the exact same time, so why does it matter if they are created at the exact same time? Ultimately, we do not really need to know exactly how God chose to create twins, when the second comes into existence or when there are two people with souls. While it might be interesting to know, our fundamental respect for the sanctity of human life is not dependent on it. We know for sure that a new, unique, individual, precious human life comes into existence at fertilization/conception. Whether or not there is a second life present then or a few days later, does not change the fact that we can never directly and intentionally kill any unborn human life.

Beginning of Life metaphor:

Oftentimes the pro-choice side uses metaphors to try to obfuscate the reality of the situation. One popular metaphor is comparing fetal development to the building of a car. At what point does a car become a car instead of just a pile of mechanical pieces? Certainly, if you own just a random steering wheel, a single tire and a rearview mirror, you do not own a car. However, if you have all the parts except for a rearview mirror, people would agree that you would have a car, just one that is missing a part. The question then becomes, "When does the car become a car, instead of just being pieces with the potential to transform into a car?" Pro-choicers then compare this to fetal development saying that the embryo/fetus in the early stages of development only has the potential to be a person. They then disagree about when exactly personhood comes about, as noted by the above differences. One problem with this metaphor is that the embryo stage is a natural stage in human development and that embryo already is fully human. Yes, they are different from adult humans, but that does not mean that they are less human.

A more realistic comparison is a polaroid picture. When one takes a picture with a polaroid camera, one has to wait for the picture to develop in order to see the image. However, if one were to rip the picture just because (s)he could not see it yet, one would ruin the image. Similarly, if one aborts an embryo/fetus and claims that since they cannot see the embryo yet, then the embryo must not exist; then one is destroying the life that is actually there. Both the picture and the embryo are originally invisible to the human eye, but both are their authentic selves from the beginning and just need time to develop.

6.

Fetal Development

Fertilization Age / Time from Conception:
(add approximately 2 weeks to get gestational age)

2 weeks: Implantation happens around 6 days (Conception), 1/100 in long (Moore 6), "composed of hundreds of cells" (Cell Division)

"Blood cells begin at 17 days" (Motion)

"The baby's eyes begin to form at 19 days" (Motion)

"By the end of the 20th day the foundation of the child's brain, spinal cord and entire nervous system will have been established" (Motion)

4 weeks: ⅙ in long, developed head, trunk and budding limbs, brain and tissues developing (Stages First Trimester), "28 days the embryo has the building blocks for 40 pairs of muscles situated from the base of its skull to the lower end of its spinal column" (Motion), "30 days the primary brain is present and the eyes, ears and nasal organs have started to form" (Motion)

5 weeks: ⅜ in long (You and Baby), "10,000 times bigger than (s)he was at conception" (Voyage Week 7 G.A.)

6 weeks: ½ in long, 4 chambered heart, fingers and toes are on the hands and feet (Stages First Trimester), muscle movements (Motion), "The primitive skeletal system has completely developed by the end of six weeks" (Motion)

"Earliest reflexes begin as early as the 42nd day" (Motion)

8 weeks: 1 ¼ in, "the beginnings of all key body parts are present," eyelids close (Stages First Trimester) "Bone cells begin to replace cartilage" (Moore 9), "The lines in the hands start to be engraved by eight weeks" (Motion)

10 weeks: 2 ½ in head to butt, ½ oz, nails, "small, random movements" (Stages First Trimester)

12 weeks: 3 ½ in head to butt, 1 ½ oz, ability to swallow, has blood and urine (Stages First Trimester) Periods of sleeping and being awake (Moore 11), "By the end of the first trimester (12th week) the fetus is a sentient moving being" (Motion)

> "The facial expressions of the baby in his third month are already similar to the facial expression of his parents" (Motion)

14 weeks: 5 to 6 in, 3 or 4 oz, hair is present on the head (Moore 12)

16 weeks: 6in, 4 ½ oz, "respond(s) to sound" (Moore 13)

18 weeks: 6 ½ in, 5 to 8 oz, mother can feel movement, baby can suck thumb and have hiccups (Moore 16)

20 weeks: 9in, 1 lb (Moore 17); Baby can move after hearing her/his father's voice (Voyage Week 22 G.A.), "The fetus responds to taste, temperature, pain, pressure, movement, and light with changes in heart rate and movement patterns" (Voyage Week 22 G.A.)

22 weeks: 12 in, 1 ½ to 2 lb, eyelashes and eyebrows can be seen (Moore 18, 19)

24 weeks: 13 in, 2 lb (Moore 20);

26 weeks: 13 ½ in, 2 ¼ lb (Moore 21), baby can "control body temperature" (Fetal Third Trimester); "The eye lids are now partially open" (If pregnant MNDH)

28 weeks: 14 in, 2 ½ lb (Moore 22), "can open and close its eyes," the lungs "are capable of breathing air" (Stages Third Trimester), "makes grasping motions" (Women's Right Texas)

30 weeks 16 in, 3 to 4 lb (Moore 23); "Skin is thicker" (If Pregnant MNDH), "Pupils will react to light" (Women's Right Texas) 32 weeks: 17.7 in (NHS UK), 4 to 5 lb, do not necessarily need to stay in the NICU (Moore 24); smooth skin and fingernails are the full length (If Pregnant MN); moving around at least 60% of the time, bones harden, breathing movements are rhythmic (Women's Right Texas)

34 weeks: 18.7 in (Week 36 GA), 5 to 6 lb, "fetus can turn and lift its head" (Moore 25); skin is not wrinkly (Women's Right Texas)

36 weeks: 19.6 in (Week 38 GA), about 6.5lb; has a firm grasp (If Pregnant MNDH)

38 weeks (40 weeks GA): 20.2 in (Week 40 GA), about 7.5 lb; "small breast buds are present" (Women's Right Texas)

7.

Abortion Procedures

Hearing the details about abortion procedures can be very emotional and heart wrenching. You might think to yourself, "I'm already prolife. Why do I need to know how it works? I'm not going to do it." However, it is still extremely important to understand how the different procedures work, so that you can be well-versed in all aspects of the issue. The new knowledge will then help you converse articulately with others and bring up specific points that are directly tied to the actual procedures. Knowledge is power. The following descriptions are straightforward, easy to follow and backed by multiple peer reviewed scientific journals. Keep in mind that when talking with others, a direct, unexaggerated, medical based explanation of abortion will suffice in unveiling its horrific nature.

Abortion Pills:

Stage in Pregnancy:

"Often, gestational age is established by ultrasonography, but sometimes history and physical examination can accurately confirm gestational age during the 1st trimester" (Casey Induced). This recommendation can be highly problematic as an ultrasound is much more precise in determining gestational age, than any knowledge of last period, guess about conception, or even physical exam. If a woman takes the abortion pills when she is much farther along because the abortionist had a wrong guesstimate of gestational age, then she can have severe complications from a half-done abortion. Sometimes the baby can even survive (which would be a beautiful miracle!)

Commonality:

"Medication abortions account for 36% of U.S. abortions below 9 weeks" (Hufbauer et al. 9).

Types:

The most common set of pills is Mifepristone, followed by Misoprostol a couple days later. It is also more effective than the less common options of methotrexate/misoprostol or misoprostol alone (Ajmal et al.).

Mifepristone / Misoprostol:

Stage of Pregnancy:

The Mifepristone/Misoprostol combination is the ACOG's "preferred therapy for medication abortion up to 70 days of gestation" (Krugh et al.).

How med students are taught to frame the narrative:

"You take one pill that stops the pregnancy from growing, then take other pills later, which will cause cramping and bleeding" (Hufbauer et al. 22).

How it actually works:

-   Mifepristone is taken at the abortion clinic. The Mifepristone blocks the progesterone's ability to bind to the appropriate receptors. This leads to a severe shortage of oxygen and nutrients for the embryo, causing the demise. (Essentially the embryo is starved and suffocated.) It also causes the cervix to soften and dilate, as well as starting the woman's contractions (Delgado et al. 4)

-   Misoprostol is then taken orally or inserted vaginally, 24 to 48 hours after mifepristone was consumed (Ajmal et al.). It "causes cramps, heavy bleeding, and expulsion of the embryo" (Abortion Consent 9).

-   Follow up visit is with the provider, seven to fourteen days from when mifepristone was first consumed to make sure that the abortion was complete and that the woman's health is not in jeopardy (Questions and Answers). While completion is most thoroughly checked by ultrasonography or hCG testing, the TEACH abortion hand book lists telemedicine as an appropriate method of follow up, although no actual physical assessment would have taken place (Hufbauer et al. 86) "Aspiration abortion might be needed if medication abortion fails or the woman bleeds heavily during the medication abortion" (Ajmal et al.).

Time:

> "Cramping and bleeding [starts] 1-4 hours after taking the miso-
> prostol" and "last(s) for several hours" (How Pill).
> "Average bleeding duration is 9 days" but can last up to 45 days
> (Hufbauer et al. 85).

Potential Side Effects:

Although the abortionists are required "to report any observed
complications," they are not required to do any follow-up appointments (Reardon et al.). Rather, they tell the women to seek care, if
needed, at an emergency room and, thereby, turn a blind eye to the
gravity and frequency of the side effects. Further, "the emergency
room staff is not required to report the complications" (Reardon et
al.). This has assumedly led to chronic underreporting of complications in the United States. That being said, "records-based study examining all Medicaid-funded abortions from 1999-2015 (surgical
n= 361,924; mifepristone n=67,706) revealed that 35.5% of women
undergoing a mifepristone- induced abortion required emergency
room treatment within 30 days of their abortions. This complication
rate was approximately 100 times higher than reported in data vol-
unteered by Planned Parenthood clinics and 273 times the AE complication rate reported to the FDA" (Reardon et al.). Similarly, "A
record linkage study of all abortions in Finland (surgical n=20,251,
medical n=22,368) found that 20.0% of women undergoing mifepristone-induced abortions experienced adverse events, including
15.6% suffering hemorrhage, 5.9% requiring surgical intervention,
and nine deaths per 100,000 abortions" (Reardon et al.).

Reported complications include:

"About 85% of patients report at least one adverse reaction following administration of MIFEPREX and misoprostol, and many
can be expected to report more than one such reaction" (7) Nausea
51-75%, Weakness 55-58%, Fever/chills 48%, Vomiting 37-48%,
Headache 41-44%, Diarrhea 18-43%, Dizziness 39-41% (MIF-
EPREX 7), Incomplete abortion (If Pregnant MNDH), Continuation of pregnancy (Clinical Handbook 33)

Mifepristone:

"Bleeding for more than 30 days" (4), "Hemorrhage", "Sepsis",
"Hospitalization" (8), "Serious bacterial infection", "Fatal septic
shock", "Blood transfusions", and "Surgical uterine evacuation occurs in about 1% of patients" (5), "Post abortal infection", "Anemia",
"Anxiety", "racing pulse, heart palpitations, heart pounding", "Syn-
cope, fainting, loss of consciousness, hypotension... light-headed-
ness", "Shortness of breath", "Back pain, leg pain", and "Uterine
rupture, ruptured ectopic pregnancy" (MIFEPREX 8). "28 reports of
deaths in patients associated with mifepristone since the product
was approved in September 2000 (Questions and Answers).

Misoprostol:

Abdominal pain, Nausea, Vomiting, Diarrhea, Fever, Chills
(Second Trimester Induction), Maternal shock, Maternal death,
Headache, Constipation, Aches/pains, Fatigue, Drowsiness, Rash,
Breast pain, Abnormal taste or vision, Deafness, Upper respiratory

tract infection, Bronchitis, Pneumonia, Hypotension and hypertension, Gastrointestinal bleeding, inflammation, and infection, Anaphylactic reaction, Urinary tract infection, Anxiety, Change in appetite, Depression, Dizziness, Thirst, Impotence, Loss of libido, Neurosis, Confusion, Sweating, Stiffness, Back pain, Anemia, Uterine rupture, and the "possibility of a general adverse effect on fertility" (Pfizer 1, 7-11), "a rare case of maternal death" (Krugh et al.)

Effectiveness:

"95% of pregnancies that have lasted 8 to 9 weeks
87 to 92% of pregnancies that have lasted 9 to 11 weeks" (Casey W.H.I.)

"95-99% (effectiveness of pills) up to 9 weeks, 91-93% up to 10 weeks" (Hufbauer et al. 22)

When Mifepristone is used without Misoprostol, there is up to a "25%(approx.) embryo or fetus survival rate" (6) and "an incomplete abortion rate of 20-40%, as determined by the end point of complete expulsion" (Delgado et al. 4). However, Misoprostol's original use was not intended for abortion, but rather for "the prevention and treatment of NSAID-induced gastric ulcers in patients at high risk for ulceration" (Krugh et al.).

Additional Notes:

-   "Mifepristone and misoprostol do not terminate ectopic pregnancy" (Clinical Handbook 28).

- "A woman is 30% more likely to die from a missed ectopic while undergoing chemical abortion than if she had not chosen an abortion" (Reardon et al.)
- "According to the FDA, the abortion pill has not been studied in women who are heavy smokers" (Methods).
- When the FDA was looking to hastily approve the Mifepristone / Misoprostol combination, they waived, "the normal requirement of two randomized, blinded, placebo-controlled trials demonstrating significant efficacy and minimal risks" and alternatively looked at "a single published trial that was non-blinded, non-randomized, and utilized only a historical, non-concurrent control" (Reardon et al.)
- "Even after widespread use for over 20 years, there have still been no randomized trials investigating the mid- to longer-term complications associated with mifepristone-induced abortions" (Reardon et al.).

Abortion Pill Reversal:

Since Mifepristone acts by blocking the progesterone receptors, an influx of progesterone acts to ensure that enough progesterone is able to bond to the appropriate receptors to insure continued fetal growth and well-being (Delgado et al. 5). The influx of progesterone in a large dose compensates for the fact that Mifepristone has a "binding affinity for the progesterone receptor (that) is 2.5-5 times that of progesterone" (Lalitkumar et al.).

Effectiveness:

"when taken in the first 72 hours of Mifepristone and before Misoprostol"

Type of Progesterone:

- Intramuscularly dose - 64% reversal rate" (Delgado et al. 8)
- High-dose oral - 68% reversal rate" (8)

Gestational Age:

5wk - 25%

6wk - 46%

7wk - 49%

8wk - 61%

9w - 77% (Delgado et al. 10)

Additional Notes:

- "There is no increased risk of birth defects in babies born after mifepristone reversal" (Delgado et al. 8).
- Maternal age did not seem to affect the rate of reversal (8).
- Check with a doctor about types of progesterone and its first pass effect.

More information and quick help can be found at https://www.abortionpillreversal.com/ and their abortion pill reversal 24/7 helpline at: 877-558-0333

Methotrexate / Misoprostol:

Stage of Pregnancy:

Less than 50 days (Protocol 1)

Steps:

- Physical and pelvic exams, lab work and recommended ultrasound (Protocol 3)
- Methotrexate is administered orally or intramuscularly (Protocol 4). "Methotrexate interferes with DNA synthesis, repair and cellular replication", thereby causing the fetal demise (Hospira 19). It then leads to "uterine contractions and expulsion of embryonic or fetal tissue" (Baird). It is effective because "fetal cells... are in general more sensitive to this effect of methotrexate", than the maternal cells (Hospira 19.
- Misoprostol is then inserted 2 to 6 days later (Protocol 4) and specifically helps with the contractions and expulsion stage.
- An ultrasound is typically done 7 days after methotrexate. More Misoprostol is administered if abortion appears to be incomplete. If incomplete at day 29-45, a vacuum aspiration abortion is done (Protocol 4).

Time:

"Abortion occurred in the 24 h(ours) following the initial or repeat misoprostol dose in 70.7%" of women (Creinin).

Potential Side Effects:

When used for abortion:

Nausea 47%, Hot flashes 43%, Diarrhea 22%, Dizziness 21%, Headache 16%, Vomiting 12%, Incomplete abortion ~5% (Creinin)

General Side Effects of Methotrexate:

Nausea, Abdominal distress, Infection, Malaise, Fatigue, Chills, Fever, Dizziness, Death, Impairment of fertility, Loss of libido, Menstrual dysfunction, Neurotoxicity, Diarrhea, Vomiting, Intestinal perforation, Pulmonary toxicity, Diabetes, Blurred vision, Hyperglycemia, Stress fracture, Convulsions, Respiratory failure, Acne, Rashes, Chromosomal damage to bone marrow cells, Benzyl alcohol toxicity, Ulcerative stomatitis, Anaphylaxis, Anemia, Increased risk for bacterial, viral and fungal infections, Possible irreversible renal and liver failure (Hospira), and Lymphoma (Hamed et al.).

Effectiveness:

It can be ">90% effective at causing abortion in women at less than 49 days' gestation" (Creinin). "The failure rate at 50-56 days is approximately twice that at 43-49 days" (Protocol 2).
Additional Notes:

- "Neither (Methotrexate nor Misoprostol) has been approved for induction of abortion" when used together, by the U.S. Food and Drug Administration, but rather are being used off-label with clinician discretion (Protocol 1, 2005).

- Methotrexate has potential to be seriously detrimental to future fertility. In general, doctors recommend a 6 month wait time for conception after being treated with Methotrexate. Even small doses can be harmful. "Where low-dose MTX was used within 1 year of pregnancy[,] a high rate of spontaneous abortion is seen" (Lloyd et al.). Even if someone uses Methotrexate before a current pregnancy, the current pregnancy could still be at risk "Because methotrexate is widely distributed and persists in the body for a prolonged period, there is a potential risk to the fetus from preconception methotrexate exposure" (Hospira 15).

Reversal:

Since the Methotrexate / Misoprostol type of abortion is less common and not even FDA approved for that specific function, information on reversal potential is much sparser (Protocol 1, 2005). However, folinic acid and its derivatives seem to be the key to success. While "methotrexate interferes with DNA synthesis, repair and cellular replication" (Hospira 19), "folic acid plays a crucial role in the synthesis of DNA and cell replication" (Rattu et al.). Because of this, "products containing folic acid or its derivatives may decrease the clinical effectiveness of methotrexate" (Hospira 11).

When specifically considering a pregnant woman who has been exposed to Methotrexate (MTX), *QJM: An International Journal of Medicine* advises that she stops taking the drug, "follow[s] counseling, (and) be offered treatment with folinic acid for a least 5 months in order to minimize the MTX effects on the fetus" (Lloyd et al.).

Preferably, she "should continue (folinic acid) supplementation throughout pregnancy" (Lloyd et al.). Folinic acid is necessary for the hopeful "reversal of the anti-folate effects of MTX" (Lloyd et al.) and is even "the preferred antidote for MTX poisoning" (Hamed et al.). Knowing this, the National Abortion Federation recommends that a woman receiving a Methotrexate abortion stop taking vitamins with folate for the week following her Methotrexate intake (Protocol 2).

Leucovorin is one of the derivatives of folic acid (Hegde and Nagalli) and works to "efficiently neutraliz(e) methotrexate effects" (Hamed et al.). It does this by "exert[ing] its effects via competitive cellular uptake" (Rattu et al.). It is so effective that it has the potential to "virtually eliminat[e] teratogenic effects" (as demonstrated in an animal study), which are effects "that ac[t] during prenatal life to produce a permanent physical or functional defect in the offspring" (Lloyd et al.).

A woman who takes Methotrexate and then wishes to reverse her abortion, should speak to a pro-life doctor immediately about the best possible treatment plan for her specific situation.

Aspiration Abortion:

Commonality:

The most common type of surgical abortion through the first 14 weeks (Schnettler)

Stage in Pregnancy:

Manual vacuum aspiration used < 9wk (Casey Induced)

How med students are taught how to explain it:

"This is quick, done with me, on an exam table with instruments inside you" (Hufbauer et al. 22).

What actually happens:

- "Insert the speculum."
- "Apply the antiseptic solution to the cervix."
- Sometimes "administer a paracervical block" (Ajmal et al.) which is not 100% effective in relieving pain (Stubblefield et al.)
- "Some use a cough technique to distract during injection" (Hufbauer et al. 49)
- Misoprostol or osmotic dilators may be applied.
- 'Place the tenaculum on the cervix. Dilate the cervix to the size of the cannula" using mechanical dilators.
- "Insert cannula"
- "The procedure is completed by aspiration of the uterus using a manual or electric vacuum and not by sharp curettage" (Ajmal et al.).
- One of the ways to tell that the abortion is complete is to notice an increase in cramping (Hufbauer et al. 61)
- "Inspect the tissue" (Clinical Handbook 43)

Time: 5 -10 minutes (Ajmal et al.).

Potential Side Effects:

- Vasovagal episode, Heavy bleeding, Perforation, Incomplete abortion, Endometritis, Continued ectopic pregnancy (Ajmal et al.), Infection, Torn cervix, Future premature birth (If Pregnant MNDH), Intrauterine adhesions (Huchon et al.)
- "higher risk of miscarriage, particularly those occurring within the first 3 months of gestation, is associated with prior first trimester induced (particularly vacuum) abortion." (Sun et al.)
- "After 8 weeks of gestation, the risk of major complications appears to rise by ~ 15 - 30% for each week of delay" (Lalitkumar et al.).

Additional Notes:

"Some providers routinely use non-sterile gloves for uterine aspiration" (Hufbauer et al. 57)

Dilation and Curettage (D&C):

Commonality:

Second most common surgical type of abortion through the first 14 weeks. Done when the simpler aspiration method is deemed not sufficient (Schnettler).

Stage of Pregnancy:

Available up through 13w 6d (first trimester) (Cooper and Menefee) (Casey Induced).

Steps:

- Bimanual exam is done (Cooper and Menefee).
- "Vaginal preparation with an antiseptic solution" (Cooper and Menefee).
- "A bivalve or weighted speculum is placed in the vagina" (Cooper and Menefee).
- Possible local anesthesia (Cooper and Menefee).
- Early dilation can be achieved by chemical agents or osmotic dilators. One common one is Laminaria, a type of seaweed (Cooper and Menefee). However, "the Society of Family Planning does not recommend any (prior) cervical preparation for first-trimester abortions unless the woman is at increased risk of complications" (Cooper and Menefee). The day of the procedure, dilation can be achieved with physical dilators that gradually increase in size until the cervix is open enough for the curette to pass through (Cooper and Menefee).
- Suction is then applied, either manually or electrically, and a cannula, a hollow tube, is used (Cooper and Menefee).
- "The curette is applied to the walls of the uterus" (Cooper and Menefee). "A gritty texture indicates complete removal of the pregnancy" (Cooper and Menefee).

Time:

10 to 15 minutes (Methods 6)

Potential Side Effects:

- Infection, Hemorrhage, Cervical lacerations, Uterine perforation, Post-op adhesions (Hucheon et al.), Painful periods,
Menstrual cycle changes, Infertility, Bowel injury (Cooper
and Menefee) Secondary infertility from "Scarring can occur
during a dilation and curettage" "that can create adhesions
inside the uterus that interfere with future pregnancies"
(Secondary Infertility).
- If prior uterine scar (such as from a c-section), "the risk of
scar inadequacy in this case increases 1.5 times" (Dikke and
Ostromenskiy).
- Death, Mortality Rate of .6 per 100,000 legal induced abortions (Cooper and Menefee)

Additional Notes:

Fentanyl is sometimes used as a pain killer during surgical abortion procedures, even those done in the first trimester (Chin et al.)

D&E (Dilation and Evacuation):

Commonality:

# of Abortions Occurring in the Second Trimester

- "Abortion during the second trimester of pregnancy accounts for 10-15% of abortions performed worldwide" (Newmann et al.)
- 150,000+ women have second trimester abortions yearly (Jones and Weitz)
- The estimated 1% of abortions that happen after 20 weeks in the U.S. equals around 10,000 up to 15,000 each year (Studnicki Late-term).
- "Twelve percent of abortions are performed in the second trimester, the majority of these by dilation and evacuation" (Stubblefield et al.)

% of Second Trimester Abortions being D&E

- D&Es were ">98% of all second-trimester abortions" (Hammond 6)
- In the United States, "second trimester abortions conducted by D&E increased from 31% in 1974 to 96% in 2005" (Lohr et al.)

Stage of Pregnancy:

>13w (Lalitkumar et al.) (D&E MI)

After 14wk (Descriptions MN)

14 to 24 wk (Casey W.H.I.)

13 - 24 weeks (Jones and Lopez)

Steps:

- Bimanual examination (Clinical Handbook 50).
- Set up an IV (Ajmal et al.).
- Previous dilation can be achieved by osmotic dilators, chemical agents, or material dilators (Ajmal et al..
- "Ultrasound should be used during the procedure to locate fetal parts, visualize instruments, and verify the completion of the procedure" (Ajmal et al.).
- Foeticide is administered either 24 hours before or right before the D&E begins. (Hammond 33). This can take place by umbilical cord transection, potassium chloride, or digoxin injection (Edelman and Kapp 29).
  - This is to avoid violating the Partial Birth Abortion Act of 2003. However, digoxin inserted via the intra amniotic route can actually leave the fetus alive in 33% of cases (Hammond 34-35).
- Speculum is inserted (Clinical Handbook 50).
- Apply antiseptic solution to cervix (Clinical Handbook 50).
- Some kind of local anesthesia or pain medicine is administered, though general anesthesia is not recommended (Clinical Handbook 27).

- "Place traction on the tenaculum to bring cervix down the vagina" (Edelman and Kapp 26)
- Using a cannula and aspirator remove amniotic fluid (Clinical Handbook 50).
- Forceps and a suction cannula are used to remove the fetus and placenta (Clinical Handbook 50) (Descriptions MN).
    - If everything is not easily removable, then misoprostol, methergine or oxytocin can be administered and the procedure attempted again in 3-4 hours (Clinical Handbook 53).
- The tissue is checked to make sure all parts have been removed, including the four limbs, thorax, spine, calvarium and placenta (Clinical Handbook 53).

Time:

Around 30 minutes (Jones and Lopez)

Potential Side Effects:

- "The abortion complication rate... increases to 50% or higher as abortions are performed in the 2nd trimester" (Coleman et al.)
- "Two-thirds of major abortion-related complications and half of abortion-related mortality occur in pregnancies terminated after 13 weeks of gestation" (Hammond 5)
- Uterus fails to contract, Heavy bleeding, Torn or cut cervix, Uterine perforation, Pelvic infection, Incomplete abortion

(If Pregnant MNDH), Blood clots (Methods 8), Shock, Fever, Bowel injury, Maternal cardiac arrest from potassium chloride (Edelman and Kapp 31), Future fertility problems including, preterm birth, low-birth weight, and miscarriage due to cervix opening (D&E MI)

Reasons for Abortion:

According to Oxford Academic, second-trimester abortions are becoming more common as prenatal screening for fetal health conditions increase in popularity (Lalitkumar et al.). Even so, aborting because of fetal abnormalities only accounts for 2% of late term abortions (Coleman et al.). "The trajectory of the peer-reviewed research literature has been obvious for decades: most late-term abortions are elective, done on healthy women with healthy fetuses, and for the same reasons given by women experiencing first trimester abortions" (Studnicki Late-term). According to Guttmacher Institute, women gave the following reasons for aborting after the first trimester: 71% not knowing about pregnancy, 62% social concerns, 48% trouble setting up appointment, 33% afraid to tell father or parents, 30% physical health concerns, 24% wanting extra time to think about things, 13% wanting a specific family size, and 8% outside pressure to not abort causing delay (Coleman et al.). Researchers also found that women aborting after the first trimester had some kind of emotional turbulence such as "(1) the abortion was more likely to have been desired by the partner, (2) the abortion was less likely to have been desired by both parties, (3) there was more pressure from someone other than the partner to abort, (4) male partners were less likely to have been informed of the abortion until afterwards, and (5) women were more inclined to have left the partner before undergoing the procedure" (Coleman et al.)

Additional Notes:

"Second-trimester patients… were more likely to report recent use of hormonal contraception than other patients" (Hammond 4)

Induction Abortion (Late Term Abortions):

Stage of Pregnancy:

Typically done after 16 weeks (Second Trimester Induction)

Commonality:

- "The procedure most commonly used for second-trimester genetic terminations seems to be labor induction" (Shulman et al.)
- Four abortion clinics publicly perform abortions after 24 weeks LMP as of spring 2022 (Kimport)

Steps:

- Cervical prep beginning about 3 days beforehand, includes numbing the cervix and using dilators and laminaria (seaweed) (Abortionclinics.org).
- Digoxin (24 hours prior) or intracardiac potassium chloride is administered (Hammond 33). It causes fetal demise and

ends the heart beat. This is to avoid violating the Partial Birth Abortion Act of 2003. However, "Intraamniotic digoxin may fail to cause demise in up to one third of cases" (Hammond 34-35)

- Misoprostol and/or pitocin start contractions and dilate the cervix (Second Trimester Induction)
- "Delivery of the fetus and placenta usually occurs within 24 hours but can be longer" (Second Trimester Induction)
- If not all the tissue is excreted, then tools will be inserted through the vagina to remove the remaining tissue from the uterus (SCDH methods).
- Hospital stay can be two or three days (Second Trimester Induction)
- Essentially, this procedure closely mirrors childbirth. The woman still goes through labor and delivery, but the fetus dies before passing through the birth canal.

Time:

"Most women who are induced deliver in 10-20 hours" (Methods SCDHEC). Delivery can occur within 24 hours, but one could stay in the hospital for 2 or 3 days. (Second Trimester Induction)

Side effects:

- Infection, Heavy bleeding, Stroke, High blood pressure, Blood clots, Torn or cut cervix, Perforation of uterine wall, Incomplete abortion, Anesthesia complications, Baby might be born alive (If Pregnant MNDH) (Auger et al.), Ongoing pregnancy, Remaining tissue (Induction) Pelvic Infection 5%, Incomplete abortion... up to 36% (Abortion Risks LA)

- Side effects of the Misoprostol medication
    - Abdominal pain, Nausea, Vomiting, Diarrhea, Fever, Chills (Second Trimester Induction)

VBAC:

Many times, VBACs (vaginal birth after cesarean) are discouraged for women having consecutive pregnancies, unless they are specifically working with a holistic minded doctor. This is because of concern about the c-section scar and how the woman's body will be able to handle labor and delivery. However, if the woman is considering an abortion instead of a live birth, these concerns seem to fly out the window. A 2019 journal analysis wrote, "Termination of pregnancy in women with a scar on the uterus involves the use of any available method" (Dikke and Ostromenskiy). In case you were confused by this and concerned about gestational age or number of previous scars, it continues, "In case of interruption of late pregnancy or foetus death in patients with the uterine scar, in most cases induction of confinement is more preferable than planned caesarean section irrespective of the number of past caesarean section operations" (Dikke and Ostromenskiy). An article in the *Contraception* journal agreed that up to 35 weeks labor induction abortion was acceptable and "history of cesarean section was not associated with higher morbidity except risk of uterine rupture" (Mazouni et al.). This exception is a huge exception! This is incredibly alarming and seems to dismiss the complexity and individuality of the situation, and a proper analysis of maternal risk factors.

Rare Hysterotomy Abortion (Very similar to a c-section) & Hysterectomy Abortion:

Stage of Pregnancy:

After 24 weeks (Methods SCDHEC)

Commonality:

From 1967 to 1973 approximately 25% of abortions were done by hysterotomy (W.U. 1973). Now, they are much rarer, but will occasionally take place because of "low lying placenta, failed induction for medical termination of pregnancy [or] previous caesarean section" (Siwatch et al. 19). In the United States during 2022, there were 65 hysterotomy/hysterectomy abortions (Abortion Surveillance 2022). However, with the *Dobbs* ruling, Wikipedia 2023 began claiming that, "Providers [have begun] performing hysterotomy abortions again" (Hysterotomy Abortion). The New England Journal of Medicine confirms this saying, "Others reported that their colleagues have resorted to using hysterotomy, a surgical incision into the uterus, because it might not be construed as an (illegal) abortion" (Arey et al.).

Steps:

"Hysterotomy is an incision opening that opens the endometrial cavity usually via abdominal route for completion of (the) procedure of abortion" (Siwatch et al. 17). Procedural wise, it is comparable to a small c-section (Kulier et al.) (Methods SCDHE) (Schnettler). It is typically done in a hospital with anesthesia and foeticide is administered beforehand so that a live delivery is avoided (The Types) (Kulier et al.) (Hysterotomy Abortion). For a hysterectomy abortion, the uterus is also removed from the woman.

Time:

Comparable to a c-section and recovery

Side Effects:

- High morbidity (Clow and Crompton 321)
- Back in 1985 when it was more common, there was a death-to-case rate of 60 per 100,000 abortions (Grimes and Schulz)
- Comparable risks as a c-section including "higher rates of blood loss, transfusion" (CS stats), scar complications (Siwatch et al.19)
    - Women should be warned about future abortions. "A scar on the uterus after caesarean section presents a high risk of abnormal attachment of the ovum and fatal bleeding during abortion" (Dikke and Ostromenskiy).
- Comparable risks to a general hysterectomy including: Infection, bleeding, blood clots, damage to other body parts,

and adverse reaction to anesthesia (Abdominal Hysterectomy)

Regular Hysterectomy v. Hysterotomy/Hysterectomy Abortions:

Banning hysterotomy abortions should not be confused with regular hysterectomies - procedures done to treat a health condition of the mother without involving a pregnancy. Hysterectomy is "the surgical removal of the uterus" and "is the most common nonobstetric surgery performed on reproductive-aged women in the United States" (Wang, James et al.). While their names are very similar, the purpose of these procedures is entirely different. Hysterotomy (or hysterectomy) abortions have as their goal the direct and intentional ending of the fetal life. Hysterectomies, with no concurrent pregnancy, have as their goal the medical well being of the woman. Even a woman who has an early c-section, places the baby in the best NICU care available, and then quickly has a hysterectomy for her own severe health reasons, is not undergoing an abortion procedure. What makes the hysterotomy/hysterectomy abortions unique is their deliberate and intentional ending of the unborn, human, fetal life.

Rare Intrauterine Instillation:

Stage of Pregnancy:

16-24 weeks (specifically for hypertonic-saline types) (Ferguson) (Lawson et al.)

Commonality:

Intrauterine instillation abortions used to be much more common. However, they "decreased from 57% to 0.4%" in the U.S. from 1974 to 2005 (Lohr et al.). In 2020 - 2022, there were a total of 478 instillation abortions (Abortion Surveillance 2020) (Abortion Surveillance 2021) (Abortion Surveillance 2022).

Steps:

- Empty bladder.
- Antiseptic solution
- Local anesthetic (Kerenyi et al.)
- Some of the amniotic fluid is removed via a needle
- It is replaced by a concentrated salt solution or a prostaglandin which is also inserted through a needle going through the abdomen.
- This induces labor.
- Delivery happens about 24 to 48 hours later.
- The fetus is (typically and supposed to be) delivered dead (Glossary NC CSCHS).
- This has typically taken place in a hospital (Ferguson).

While it is possible to perform an extra-amniotic hypertonic saline-induced abortion where the solution is inserted through a device placed by the cervix, "in the United States, the saline method has been performed almost exclusively by transabdominal instillation of the hypertonic solution directly into the amniotic sac" (Kerenyi et al.) (Ferguson).

Sometimes urea would be used instead of saline (Edström 485).

Time:

24 to 48 hours from administration (Glossary NC CSCHS)

Side Effects:

Retained placenta, Hemorrhage, Fever, Incomplete abortion (Piya-Anant and Swasdimongkol), Death, Shock, Brain Damage, Convulsions, Solution entering the maternal circulation (Edstrom and Odar-Cederlof), Vomiting (Embrey and Hillier 589)

Specifically with the hypertonic-saline solution:

First attempt unsuccessful, fever, irregular menstruation, intra-uterine adhesions, incompetent cervix (Borenstein et al.), Salt overload, Water intoxication, Severe bleeding, Seizures, Coma (Ferguson), Live fetal birth, maternal death (Kerenyi et al.)

Additional Notes:

When specifically discussing the risk of live birth using intra-amniotic hypertonic saline instillation abortions, Dr. Piya-Anant remarked, "Live birth makes psychiatric trauma for patients and medical personals… especially for late terminations without fetal malformations when according to local guidelines one has to resuscitate" (Piya-Anant and Swasdimongkol). These live births speak volumes for the humanity and pure personhood of these tiny children.

It is tragic that there are multiple late term abortions being completed on perfectly healthy children and that when they are born alive, the doctors are annoyed by having to resuscitate them instead of being in awe of the beauty and sanctity of their life.

Selective Fetal Reduction:

Selective fetal reduction is choosing to abort one or more fetuses in a pregnancy with multiples. With twins, for example, one twin would be aborted and one would not. This, commonly, happens when in-vitro fertilization (IVF) has been used. IVF "is the single biggest contributor to multiple pregnancy. Following IVF conception, one in five births results in multiple pregnancy" (Beriwal et al.). The rate of IVF twins, 23%, is remarkably higher than the rate, 2%, of twins occurring from natural conception (Sam et al.). In IVF, multiple embryos are oftentimes entered into the woman, in hopes of at least one implanting. "Many physicians or centers [do] not offer [single embryo transfer] due to their pursuit of a successful rate in one transfer cycle" (Wang, Chao et al.). If multiple embryos implant, then the doctors and parents choose which one(s) to reduce based on health, location or gender. This practice started during the 1980s and is now considered commonplace and "an integral fail-safe of infertility practice" (Evans et al.).

Stage of Pregnancy:

Often done in the second trimester.

"There is (an) increased possibility of spontaneous death of one or more embryos before 11 weeks of gestation, which may render the procedure unnecessary" (Beriwal et al.). "The majority of [selective abortions] (in this study) were carried out at 13-19 weeks gestation" (Sam et al.).

Commonality:

In the United States, approximately 50% of IVF babies are from a pregnancy of multiples (Evans et al. 2014). Even when the multiples pregnancy is just twins, many women still choose selective fetal reduction (abortion). Doctors have admitted that around 30% of their fetal reduction cases happen reducing twins to a single (Evans et al.). A different study had a rate of 83% (Sam et al.). This is shocking because successfully and healthfully carrying and birthing twins is very commonplace with today's advances in medicine. One must assume that non-medical motives are the driving factor behind these procedures.

Methods:

The selective fetal reduction (abortion) provider will often perform CVS testing to check for chromosomal and genetic abnormalities the day before the procedure (Evans et al.). The abortion can then take place via several methods. Once the fetus has died, (s)he remains in the uterus for the remainder of the pregnancy (Wang, Chao et al.). If selective fetal reduction (abortion) is done in the third trimester, then the women will deliver that baby as a stillborn (Beriwal et al.).

Methods for fetal demise include:

"**Intracardiac injection of potassium chloride** (KCI) was the most prevalent method (75%)" (Sam et al.). A needle is injected into the fetal heart and KCI is dispensed (Beriwal et al.).

Radio frequency ablation occurs between weeks 15 and 27. A needle goes into the fetal abdomen, adjacent to the umbilical cord, and a radiofrequency energy heats up to 110 degrees C for a duration of 3 minutes. This may be repeated two or three times (Beriwal et al.) (Liu and Khalil). "KCI injection could result in the death or neurological impairment of the other fetus(es) sharing the placenta" (Liu and Khalil).

Bipolar diathermy cord coagulation is usually done between weeks 18 and 27. It involves a skin incision leading to an intra-amniotic port. Ultrasound and local anesthesia are used. Forceps are then used to grab the umbilical cord. The cord is then coagulated (solidified) in several areas by bipolar electrocautery, preventing the blood from being able to flow through the umbilical cord. Each point of coagulation takes about a minute (Beriwal et al.). Approximately an 18% chance that the other fetus also dies, a 21.9% chance for premature rupture of membranes, and a 44% chance of preterm birth (Liu and Khalil). There is a higher risk of future morbidity and possible death of desired fetus, possible "preterm prelabour rupture of membranes" preterm birth or miscarriage "associated with residual dead fetoplacental tissue at mid-gestation" (Beriwal et al.). Sometimes the death of the co-twin is after (s)he is born (Beriwal et al.).

Fetoscopic laser ablation occurs when the fetoscope enters the uterus and then "laser coagulation is used to ablate the umbilical cord of the fetus" (Liu and Khalil). The fetus dies from a lack of oxygen and nutrients.

Intrafetal laser ablation takes place by injecting a needle, followed by laser fiber, into the fetal abdomen. Laser is then used for coagulation, causing the blood to stop flowing. Fetal death is confirmed approximately one hour later. It "had a co-twin death rate of 46% within 2 weeks following the procedure" (Liu and Khalil).

Microwave ablation is a newer method and works by inserting a coaxial antenna into the fetal abdomen, which then gives off microwave energies for approximately 3 minutes (Liu and Khalil).

Other methods include "lignocaine, extra amniotic prostaglandin/urea/saline, intraamniotic urea, (and) umbilical KCL" (Sam et al.).

Additional Potential Side Effects on Desired/Remaining Fetus:

Death, Physical impairment, Aborting the "wrong" one, (Beriwal et al.), Low blood flow, Cerebral injury (Beriwal et al.), Premature birth, Low birth weight (Cheang et al.), "Increased the miscarriage rate and reduced the live birth rate" (Wang, Chao et al.)

Additional Potential Emotional Side Effects on Mother:

"30-70% experience acute feelings of anxiety, stress, and emotional trauma" (Beriwal et al.), "Feelings of guilt, grief and mourning, emotional pain, depressive reactions, fear" and "regret" (Bergh et al.), "Long-term psychological impacts" (Liu and Khalil), A year later, "one-third of mothers still suffered from depressive symptoms of sadness and guilt" (Liu and Khalil).

Placating Parental Preferences:

If there are such great risks associated with selective fetal reduction, then why is it so common? Well, in addition to wanting to control the sex and health of the children that they will give birth to, expectant mothers are loaded down with excessive scare tactics. The IVF doctors threaten the risks of cerebral palsy, asthma, poor vision, poor motor skills and a lower IQ for the remaining child, if reduction does not take place (Evans et al.). The mothers and fathers are cornered into the position of feeling like selective fetal reduction is the only way to have a healthy and "normal" child. In this way, the doctors try to spin it as a positive proactive choice for their future child, instead of a very grave negative choice against their unborn child. If that were not enough, the doctors claim that both/all the fetuses are at risk of mortality and they also extrapolate on maternal risks. "The risks associated with multifetal pregnancies include increased perinatal mortality and morbidity, maternal pregnancy complications, emotional distress such as fatigue, stress and depression and socio-economic problems" (Bergh et al.). Clearly, this is large undue pressure to reduce, at a time when the mother is likely already feeling hormonal, sick and extra emotional.

That being said, the above quotations are frequently used to justify reducing a multifetal pregnancy even when the main cause of such a reduction is due to preference. The Fetal Diagnosis and Therapy journal admits that, "In recent years,... the vast majority of such cases (of fetal reduction) are medically less complicated" (Evans et al.). It continues, "For patients reducing to twins, the overwhelming preference is for one of each (gender)" (Evans et al.). Thus, what was originally supposedly trying to create a better quality of life for the remaining child, can no longer hide even under that poorly orchestrated defense.

Another recent phenomena is gay men using IVF to father two babies, one with sperm from each man. In cases where multiple embryos have implanted in the surrogate mother, the embryos can be tested for paternity and then reduced accordingly. The *Fetal Diagnosis and Therapy Journal* recounts the stories where, "The couples desired FR for the usual clinical reasons, but they requested if possible to be left with twins - one fathered by each of them. We chose to consider this request in the same vein as gender preference" (Evans et al.). The preferences do not end there. "Today patients in the United States can choose donors based not only on their height, hair color and ethnicity but also on their academic and athletic accomplishments, temperament, hairiness and even the length of a donor's eyelashes" (Padawer). TikTok is overflowing with examples of people publicly praising their IVF designer babies who have the "perfectly" concocted body. Ultimately, selective fetal reduction is geared towards placating parents instead of protecting the preborn.

This is the tragic consequence of devoiding morality from the reality of the situation. Supporters of fetal reduction remark, "Compromise is often the right path because all moral judgments are fallible and their contours change as circumstances change" (Evans et al.). This statement could not be more incorrect. Morality is not relativistic. Whether one chooses to acknowledge it or not, the sanctity and dignity of each unique human life is an absolute fact.

In helping people acknowledge this fact, ultrasound is a key tool. It's been noted that, "[a couple's] observation of fetal movements on ultrasound examination supported their decision to decline reduction" (Bergh et al.). This is an important reminder that highlighting the reality of the situation, such as two beating hearts, also highlights the moral weight of the situation and can lead to a parental change of heart.

Intact D&X (Federally banned in the U.S.)

The Intact D&X abortion procedure was banned in the United States at the federal level by the Partial Birth Abortion Ban Act of 2003 because of its excessively "gruesome and inhumane" nature (United States 2003). Before the ban, D&X procedures were performed after 19 weeks and had many serious risk factors (*Stenberg*). The dissenting judges in *Stenberg v. Carhart* describe the procedure as, "breech extraction of the body of a live fetus, excepting the head… scissors are inserted in the back of the head, the fetus' body, wholly outside the woman's body and alive, reacts as though startled and goes limp", a vacuum is used and "the heart of the fetus may continue to beat for minutes after the contents of the skull are vacuumed out" (*Stenberg* Opinion 959, 960, 963).

Telemedicine:

Telemedicine health has been a growing phenomena ever since COVID-19 shut down many in person doctor visits. Even after in-person visits became available again, many people prefer telemedicine for the convenience factor. It certainly did not take the pro-choice groups long to capitalize on this trend and increase easy access to abortion pills. Post *Dobbs v. Jackson* telemedicine abortions have increased further, "137% from April 2022 to December 2022" (Ranji and Diep).

Groups and websites like PLAN C, guide women in finding abortion pills online, for at-home abortions, and even go so far as to advertise "pills in advance" (Plan C). Essentially, one can order abortion pills and then keep them on hand in case one gets pregnant in the future. This leads to a plethora of concerns about people stockpiling pills, giving them to underage or unqualified individuals, overdosing and many more problems. It is alarming to think of the grave harm and manipulation of women that could easily occur from these pills getting in the hands of pimps and sex traffickers. Abortion is already riddled with side effects and risks. These only increase exponentially if women are pressured into taking them in secret and devoid of proper medical contact. Further, Plan C also advertises "websites that sell pills... generic medications (from India)" and that there is "potential legal risk" (Plan C). There are so, so, so many legal and physical risks with such a scheme.

Part of the risky nature of this is due to the fact that, "Currently, no established definition exists of what constitutes 'medical abortion through [telemedicine]'" (Endler et al. 2, 2020). The Kaiser Family Foundation admits that the patient usually fills out an online form and then is sent the medicine in the mail. No exam, bloodwork, ultrasound or confirmation of dating is required (Ranji and Diep). Aside from the unborn life, this is dangerous for the woman since abortion pills lose effectiveness as gestational age increases, underlying maternal medical conditions can cause significant side effects with an abortion, and there is no way to be sure that the abortion was complete. An ectopic pregnancy or a pregnancy of further gestation could continue to grow. *The Cochrane Handbook* further points out that telemedicine "could hypothetically affect both the experience of the counseling and the adherence to the recommended dose and timing regimens which in turn could affect the safety, effectiveness and acceptability of this abortion model" (Endler et al. 2). WoW is one such telemedicine abortion group that has facilitated medical abortion to over 75,000 women around the world since inception." "WoW operates outside of the formal health-care system and provides no auxiliary clinical exams" (Endler et al. 3).

Despite the glaring concerns that come up, abortion advocates still adamantly claim that telemedicine abortion is a great and necessary advancement in patient care. When trying to increase abortion access for college students, the California Senate Bill 24 claimed, "The National Academies of Sciences, Engineering, and Medicine have found that prescribing abortion by medication techniques is no different from prescribing other medications, and have also found that the risks of providing abortion by medication techniques, including via telehealth, are low and similar to the risks of serious adverse effects of taking commonly used prescription and over-the-counter medications" (SB-24). Clearly, this is an absurd and unfounded claim. However, even if one chooses to ignore all of the negative maternal risk factors of telemedicine abortion, it is hard to find any legitimate, well-documented positive claims about outcomes. "Studies on telemedicine often include only self-reported outcome data however and data quality suffers from a high loss to follow-up and the absence of a comparison group" (Endler et al. 2). Thus, even if one completely ignores the morality of the issue and the sacredness of the unborn life, it would be nearly impossible to make any sort of credible claim that telemedicine is beneficial for women.

Cost:

Far too often, pro-choicers declare abortion to be a financially savvy solution for low income women. But, how cheap is this "cheap" option? The answer depends on how far along the pregnancy is and what kind of procedure is done.

Medication abortion (pills): up to $800 (Attia), $568 average in 2021 (Abortion Dashboard)

Surgical first trimester: up to $800 (Attia), $625 average in 2021 (Abortion Dashboard)

Surgical second trimester: $715 to $2,000 (Attia) up to $3,350 (McCann)

Abortion 20wk: median was $1,670; up to $5,386 in 2017 (Witwer et al.)

Selective Fetal Reduction: $6,500 anecdotal (Padawer)

Late term: Ranging from "a few thousand dollars to over $25,000" (Kimport)

Additional related costs include traveling out of state - hotel, transportation, and food, - leave from work, child care for previous children, follow up health care, and counseling. These costs quickly add up and can add up to hundreds of additional dollars.

In addition to the monetary cost of abortion, abortion takes a deeply personal toll on each woman, affecting her physically, emotionally, mentally, psychologically, and spiritually. While some women might feel an immediate sense of relief following the procedure, oftentimes the weight of these other factors sets in over time (Coleman et al.).

The monetary cost of abortion actually goes far beyond just the individual and greatly affects the whole country. Abortion and post-abortion care costs the United States approximately $134 million each year (Soleimani Movahed et al.)! When you expand that number to consider the number of lives lost, ~630,000, and the effect on the United State's economy; the Joint Economic Committee Republicans found that there was a cost of $6.9 trillion in the year 2019. This would equal 32 percent of GDP (Lee Economic 1). Oftentimes, pro-choicers bring up the economic impact of maternity leave and the "earnings loss mothers would be expected to incur when having a child... (but the) economic cost of abortion due (to) the loss of unborn lives is 425 times larger" (Lee Economic 1). 425 times larger! This has a massive impact on the economy. A few of the most obvious ways that it affects the economy is that abortion "shrinks the labor force, stunts innovation, and limits economic growth. It also weakens the solvency of social insurance programs" (Lee Economic 1). All of these things add up to unimaginable amounts of money. Abortion is quite literally killing our economy.

PTSD:

Although the ethics and morality of abortion has been trivialized by fanatic pro-abortion companies who seek to fund their extravagant lifestyles off of the backs of women in need, abortion is still of great moral and ethical weight. While companies can turn a blind eye to the aftermath of abortion, these women are unable to escape the reality of the tragedy that just happened. According to one study, 52.5% of women who had early term abortions and 67.4% of women who had late term abortions had symptoms of PTSD, between 12% to 20% of whom met the full diagnostic criteria (Coleman et al.). While society is belittling the emotions that accompany abortion, women who go through with the procedures are having nightmares, flashbacks, insomnia, "interpersonal sensitivity, paranoid ideation, phobic anxiety," "somatization, hostility, and psychoticism", particularly after multiple abortions (Coleman et al.). Pro-choice lobbyists attempt to deny these documented side effects, oftentimes citing the Turnaway Study. While this study has been heralded as proof that abortion is necessary for the mental well being of women, a deeper analysis of the study reveals many flaws, including "using an unrepresentative, highly biased sample and misleading questions", 68% of women refusing to be interviewed and half of the remaining women dropping out, atypical participant group, lack of proper control group, lack of knowledge of participants' prior medical history, deliberate misleading of readers, "withholding data", "exaggerated conclusions", "poor methodology and superficial analyses" (Reardon). In contrast, "the Christchurch studies, which have the most extensive preabortion history data…, revealed that abortion is significantly associated with increased rates of suicidal tendencies, substance abuse, depression, anxiety, and the total number of mental health problems" (Reardon). These correlations are very concerning and women are owed a chance to share their struggles and side effects, instead of being silenced by the pro-abortion community who wants to further perpetuate their propaganda.

Breast Cancer and Previous Abortion:

In addition to ignoring the serious mental health risk factors of post-abortive women, many pro-choice leaning scientists and media personalities also turn a blind eye to the multitude of physical health risk factors, particularly an increased risk for developing breast cancer in the future. Many of the heavily referenced studies, claiming to show no link between abortion and later breast cancer, have numerous flaws; such as selection bias, lack of differentiation between induced abortion and spontaneous abortion (miscarriage), lack of documentation about prior reproductive history and breastfeeding future children, including post-abortive women who were not old enough to be greatly at risk for breast cancer yet, and small sample size (Anderson et al.) (Guo et al.). Sometimes, the meta-analyses will simply choose to eliminate and not consider studies which have shown a clear connection between the two (Beral et al.).

On the other hand, numerous studies, conducted over a large span of years, countries, and demographics, have shown a very worrisome trend between abortion (before 32 weeks) and higher rates of future breast cancer. Interestingly, women who miscarry in the first trimester are not found to be at the same increased risk of miscarriage, since their hormone levels are usually lower (Anderson et al.).

- A 1994 study found that minor women who aborted after 8 weeks increased their risk of breast cancer by 800% (Daling et al.).
- A 2014 study of nine different European nations found that those who had the highest breast cancer rates also had the highest abortion rates (Schneider et al.).
- A 2018 meta-analysis found a 291% increased risk of breast cancer in women who had previously aborted (Brind et al.).
- Research in India and China has shown an increased risk range between six to seventeen fold (Kamath et al.) (Yuan et al.).
- In 2018, the Breast Cancer Prevention Institute found that "60 of 76 studies worldwide which differentiated between [induced abortion] and [miscarriage], showed an association between [induced abortion] and an increased risk of breast cancer, with 36 of those studies reaching statistical significance" (Anderson et al.)

Remains and Baby Remembrances:

The late LeRoy Carhart was one of the most infamous abortion doctors, specializing in abortions during the 2nd and 3rd trimester. Surprisingly, his office did not even attempt to dehumanize the fetus and referred to them as babies throughout their information brochure. They recognized the humanity of the baby so much so that their office offered remembrance boxes that were a "token of the precious time [you] and your baby had together" (Abortionclinics.org). Their listed services included: "Viewing your baby after the delivery / Holding your baby after the delivery / Photographs of your baby / Cremation services referral / Funeral arrangements referral / Footprints / Spiritual and ceremonial accommodations / Remembrance certificate" (Abortionclinics.org). It is truly fascinating how much they recognize the baby as a baby and that there is an appropriate grieving process with the loss of life but still choose to actively end the life anyway.

Much more frequently, the bodily remains of the babies are inhumanely disrespected. In Britain and Oregon, it has been revealed that the bodies were labeled as medical waste after the abortion and then burned to produce energy / heat (Burton Brown). Other locations have been reprimanded for illegally having "storage in a refrigerator", "dumping landfills or bagging and tossing in dumpsters", and stockpiling in "freezers or closets" (Burton Brown). Even methods that have been legal in the recent past are equally disturbing. Abortion clinics have been "flushing fetal body parts through the garbage disposal leading into the sewer system; (and) dumping fetal remains into an auger along with medical waste and grinding them together" (Burton Brown). Some laws did not even acknowledge the bodies, if the baby had been under 20 weeks (Burton Brown)! This led to even greater mistreatment of the human remains. Commonplace abortion has caused thousands of bodies each week to be disposed of (Burton Brown). All of these methods are glaring examples of gross mistreatment. Another common occurrence is for the human fetal tissue, "defined as tissue or cells obtained from a dead human embryo or fetus after a spontaneous or induced abortion or stillbirth", to be used in many different types of research (NIH

Grants). One of the ways in which it can be used is by "animal models incorporating [human fetal tissue] from elective abortions" (NIH Grants). It is an utter tragedy that beautiful, innocent, and precious human beings are being killed and their earthly bodies are being manipulated into animal studies.

Providers and Clinics:

While the pro-choice side will have you believe that abortion is extremely commonplace and ordinary medical care, the statistics reveal otherwise. "93% of obstetrician-gynecologists never perform abortions-at any stage of pregnancy" (Brief 2021 2). Of the providers who are performing abortions, "95% offer abortion at 8 weeks, 34% to 20 weeks, and 16% to 24 weeks" (Hufbauer et al. 9). If abortion is such a regular part of women's health care, then how could 93% of OB/GYNs stay in practice while refusing to perform abortions? A trendy new option is to have an abortion doula accompany a woman through the procedure. Birth doulas typically have received some form of special training and certification, but for abortion doulas the "field isn't licensed or regulated" (Kaur). This can lead to a false sense of medical security and potentially medical misinformation from people who lack the proper credentials. Aside from the abortion doulas, the lack of advanced medical certification for the actual abortionists is appalling. Only "1 state requires the clinician to be either a board-certified obstetrician- gynecologist or eligible for certification" and only "2 states require that providers have admitting privileges," themselves (Targeted Regulation 2023). This is alarming when one takes into account the gravity of abortion and the associated risks. Even though there are so many potential risks involved, only "28 states require providers to report post abortion complications" (Abortion Reporting 2023). This is outrageously low and puts the health, well being, and even lives of their patients at risk. It also skews stats, making it nearly impossible for women to make informed decisions about the procedures and their actual risk for complications.

Concerning abortion clinics, "almost all clinics provided abortions between 6 and 9 weeks... 64% did at 11 weeks," and 22% did "at 20 weeks gestation" in 2017 (Witwer et al.). In regard to safety precautions, only eight states have/had distance requirements for the clinic to be close to a hospital (Targeted Regulation 2023). Pro-choice Guttmacher Institute goes so far as to complain that, "17 states have onerous licensing standards many of which are comparable or equivalent to the state's licensing standards for ambulatory surgical centers" (Targeted Regulation). What are these "onerous" standards? ASCs are usually "accredited by the major independent healthcare rating agencies. ASCs are also required to maintain a strict sanitary environment; every ASC must establish and maintain programs and procedures for preventing infections. Similarly, ASCs must conduct regular, comprehensive assessments of the quality of care they provide to their patients" (Quality and Safety). Sounds like it would be in the best interest of women for abortion clinics to meet these cleanliness standards. Why would this be a problem? Overall, how can the abortion activists claim to have the women's best interests at heart when they are not providing them with the highest quality of medical practitioner and facility?

Numbers and Demographics:

Globally, every year, there are approximately 211 million pregnancies, 46 million of which result in abortion (Soleimani Movahed et al.). This means that more than 1 in 5 pregnancies is aborted. Of the pro-choice leaning countries that do not place limitations on the reasons for abortion, >75% ban abortion past 12 weeks and 88% ban it after 20 weeks (Baglini). Comparatively, this shows that the United States' policies are rather lenient and much more permissible than many other pro-choice countries.

During recent years, in the United States, approximately 1.21 million abortions occur every year (Biggs et al.). This number is estimated because the U.S. does not require abortion reporting at the national level (Sajadi-Ernazarova and Martinez). Therefore, it is up to the individual states to voluntarily report. "Three states (California, Maryland, and New Hampshire) do not report at all, and they account for 20% of the total US abortions" (Studnicki Late-Term). Back in 1985 when reporting was available, California accounted for the largest amount of abortions (Lawson et al.).

Post *Dobbs v. Jackson*, many states have enacted strict abortion regulations, causing a steep decline in number. "States that have implemented abortion bans saw a drop in the number of abortions, from an estimated 7,500 abortions in April 2022 to fewer than 10 (clinical) abortions per month after August 2022" (Ranji and Diep). As of December 20, 2024, 12 states have enacted a (nearly) total ban, while an additional 29 states have restrictions based on how far along the pregnancy is. This leaves nine states, and D.C., with no limit based on gestational age (Abortion Dashboard). Consequently, the current abortion numbers and trends are difficult to document at the present time because they are continuing to change rapidly.

Age: Unsurprisingly, teenagers had the highest abortion ratio in 2022 of any age category, totaling 50,543 abortions (Abortion Surveillance 2022). As we will discuss further, these teenagers should be given so much care and love to help them through this challenging time. Extra support from family, teachers and medical staff is all very necessary and good. With that in mind, parental involvement and consent laws are very good and can lead to a decrease in abortion as well as potentially "a reduction in the pregnancy rate of about 6%" (Dobson). While 36 states require some form of parental involvement, 35 states also have an option for a judicial bypass whereby the minor can avoid her parents after all (Parental Involvement 2023). Additionally, only two states have a requirement that the minor needs to have identification (Parental Involvement). This begs the question of how many minors are actually getting abortions, and doing so without parental consent, because they evade identification (Parental Involvement).

Marital Status: Correspondingly, 87.7% of women who had an abortion in 2022 were unmarried (Abortion Surveillance 2022). This further highlights the correlation of strong, happy marriages and families and women who feel empowered in their stage of life and supported in their pregnancies and parenting.

Previous Birth: On the other hand, it is interesting to note that 59.4% of women who had an abortion in 2022 had already given birth (Abortion Surveillance 2022). This means that the majority of women going through with an abortion had previously experienced the miracle of birth and had witnessed first hand the developing stages of pregnancy and the humanity of the child. Yet, they chose to go through with the abortion anyways. This indicates that the pro-life movement not only has to "prove" the humanity of the child, but also has to help make tangible solutions accessible to these women.

Previous Abortions: Even more astonishing is that, again in 2022, 37,216 abortions, equaling 8.2% of all abortions, were done on women who have already had three or more abortions (Abortion Surveillance 2022). These women, on their fourth or more abortion, had basically turned this gruesome, highly controversial procedure into a habit. After multiple abortions, the procedure has essentially become a mode of back up birth control.

Race: Sadly, the abortion movement was founded on racist attitudes which have since trickled down and still greatly affect our communities today. Margaret Sanger, founder of Planned Parenthood, talked at the women's branch of the Ku Klux Klan and publicly spoke about that interaction later (Infante). Now Planned Parenthood continues to covertly seek the elimination of Black Americans by their strategic center locations. "79% of Planned Parenthood surgical abortion facilities are within walking distance (2 miles) of relatively high Black and/or Hispanic/Latina populations," based on the 2010 Census data (Parker et al.). It is impossible for this to happen by random chance and is a clear sign of how they are trying to infiltrate these communities and then exploit them for profit. Tragically, it is a very successful strategy as they have prompted so many Black women to have abortions that now "Abortion is the leading cause of death for African Americans, more than all other causes combined, including AIDS, violent crimes, accidents, cancer and heart disease" (Parker et al.). In 2022, Black women had at least 134,308 abortions (163,863 in 2021), Hispanics had 72,241 abortions (85,983 in 2021) and Non-White women had 24,932 abortions (Abortion Surveillance 2021) (Abortion Surveillance 2022). Keep in mind that these numbers are likely low due to underreporting. Comparatively, "Black women have abortions at a rate of 23.8 per 1,000 women, nearly four times the rate at which White women have abortions" (Lee Economic Cost 7). As the Joint Economic Committee of Republicans so accurately points out, "This disparity results in a U.S. population that is less racially and ethnically diverse than it would otherwise be if abortions were restricted" (Lee Economic Cost 8). We have lost the contributions of so many wonderful people because of this new form of discrimination.

Religion: One factor that does not seem to play a large scale role in abortion access is religion. "Of patients obtaining abortions 37% identify as Protestant and 28% as Catholic" (Hufbauer et al. 9). Similarly, "68% of men whose partner/ spouse has had an abortion indicate their religious preference is Christian" (Care Net 42). People from all faith backgrounds are getting abortions, which is why it is so necessary to be able to argue against abortion from a moral perspective, general ethics perspective, philosophical perspective, and scientific perspective. When you use a combination of these, then you can best create the most effective explanation for the specific person to whom you are talking.

LGBTQ+: Lastly, the LGBTQ+ population is becoming increasingly present and vocal about abortion access. Now "23% of clinics providing abortions offer transgender care" (NIH cost 2). There is frequent overlap of clientele with an estimation that "LGBTQ+ people... make up as many as 16% of US abortion patients" (Chiu et al.). With the exception of lesbians, members of the LGBTQ group were more likely than heterosexual women to have an abortion, with bi-sexual women being three times more likely (Charlton et al.). In a study of "transgender, non-binary, and gender expansive adults" 31% of their pregnancies ended in abortion (Fact Sheet LGBTQ+). In total, "As many as 149,000 individuals who obtained abortion care that year identified as something other than cisgender and/or heterosexual", many of whom were on the younger age end (Chiu et al.). Clearly, the partnership between abortion access and LGBTQ "affirming care" is of notable proportions, particularly in how they advertise to the youth.

8.

Common Ground

Finding common ground with a pro-choicer might seem like it's the antithesis of your goal, but it is actually an extremely helpful tool in persuading the other person to become more pro-life. Firstly, remember the old saying, "You'll catch more flies with honey than vinegar!" This could not be more true when it comes to abortion debates and conversations. These talks are stereotypically contentious and negative, so if you can show the other person that you are friendly and interested in what they are saying, then they will be much more engaged in the conversation and willing to continue it. Secondly, pinpointing common ground at the beginning saves you a lot of time and energy, discussing pro-life talking points that they might already agree with you upon. When you find common ground you can use it as a bridge upon which to build more common ground. The goal is to keep building so much common ground that they eventually completely agree with you! Common ground provides a great starting point for the conversation, but it is also really helpful to continually be finding common ground and pointing it out to the other person. It is a gesture of friendliness and respect. To clarify, you never want to sacrifice your pro-life convictions in order to find common ground. So, the amount and type of common ground that you find will be dependent on how hard core their pro-choice stance is. Even if the common ground is really hard to find and rather small, still use it to your advantage.

Examples:

Third trimester abortion is wrong.

Babies who survive a failed abortion should be given medical care.

Abortion based on race or gender is wrong.

We both value women's rights.

It is not pro-woman to have abortion clinics not meet basic health care codes.

Adoption is not worse than abortion.

Abortion costs should not be covered by taxes.

Pro-life OB/GYN doctors in training should not have to assist abortions.

  - "Under current California law (2019), all residency programs in obstetrics and gynecology include training in abortion" (SB-24)

Basic Apologetics:

Overarching Premises and Conclusion:

An individual, unique, human being is brought into existence at the moment of conception when the sperm fertilizes the egg. Abortion is the direct and intentional killing of an embryo/fetus/baby in utero. Everyone has a fundamental right to life from the very beginning of their existence. Therefore, since everyone has the right to life from the beginning, life begins at conception, and abortion ends one's life, abortion is gravely wrong and should be made illegal.

When simplified to its core, this is what the issue of abortion is all about - respecting and protecting human life. All of the other factors, such as finances, health or even rape, can complicate an individual situation, but none of them can change the harsh reality of what an abortion really is. Knowing this, it is best to be able to respond to each complicated pro-choice argument with thorough and thoughtful answers, highlighting our care for the mothers and expounding upon common ground.

9.

Overpopulation / Global Warming

Common Ground Possibilities:

- We do have a responsibility to care for the earth, our common home.
- We should be good stewards of the earth and our environment.

Response:

While it is true that we have a responsibility to protect our environment, we have an even greater responsibility to protect the other members of the human race. This includes the unborn members who are already in existence and are fully alive. We do not take into account every environmental factor and every single person's opinion on our other actions, so why should they have such influence on whether or not a woman continues to carry a single pregnancy to term? Should a woman really surrender her power to the world and let strangers dictate whether or not she can have one baby? Keep in mind, the baby has already been created and is part way through the pregnancy. Obviously, we do not just kill other people who are adding to the proposed overpopulation problem. So why would we kill these tiniest members of the human race?

That being said, abortion, particularly the chemicals in the pills, actually creates many environmental problems. Based on the number of chemical abortions that happen in a year and how many ounces of tissue are excreted each time, it has been calculated that, "Every year, 30+ tons of chemically-tainted medical waste - human tissue, placenta, and blood - are flushed into U.S. water systems due to Chemical Abortion Pills" (Think Before). This is extremely concerning and dangerous for humans, animals, and plants. Further, it has been found that, "The metabolites used to end pregnancies remain active and unfiltered by water treatments," and thereby expose unwitting Americans (Think Before). Plants deserve better. Animals deserve better. Most importantly, people deserve better than abortion.

10.

My Body, My Choice

Common Ground Possibilities:

- We both agree that women should have equal rights.
- We both agree that women should have the ability to make medical choices when it only pertains to her specific body.
- We both want women to have control of their health care and receive excellent health care.
- We both want women to be empowered when choosing the direction of their lives.

Response:

After sharing some form of common ground and reaffirming the dignity, worth, and freedom women should rightly have, begin by pointing out that the fetus is not a part of her biological body. As simple as it is, this key distinction makes a world of difference (and is impossible to refute)! From the very moment of fertilization, that unborn child has a unique set of DNA that is inherently different from the mother's. Again, time and the natural progression of events will allow this child to grow from a single celled organism to a recognizably human baby. While it is good for a woman to have control of her own health, that control and choice ends where a new human being's rights begin. From the moment of fertilization, the unborn human being has a right to life. When a woman has an abortion, she is not simply choosing a medical procedure for herself, but rather is actively and directly ending the life of another human being. Her right to decide personal healthcare does not have the power to dictate whether or not her tiny baby lives or dies.

Your conversationalist might agree that biologically the fetus is a different human, but that since the woman's body is so greatly affected by the pregnancy, her rights over her body outweigh the fetus's rights to his/her body. This is an alarming position to take, because it, all of a sudden, starts ranking people's worth. It is eerily similar to the egregious three fifths compromise. We also need to remember that while, yes, we do have many rights over our own bodies, our bodies were not solely intended for our own glorification and gratification. Rather, the gift of our bodies enables us to make sacrifices for the benefit and well-being of others and the glory of God.

Perhaps, the most strange argument for abortion based on autonomy is one based on the abortionist's autonomy, not even the woman's. In 2022, Dr. Lauren Thaxton spoke in an interview about new restrictions and how, "the constraints on physicians' autonomy to practice evidence-based medicine have created concern about the law's long-term consequences for the medical field" and the "toll on clinicians mental health" (Arey et al.). This is wildly far fetched. It is most definitely within the rightful realm of elected government officials to place restrictions on abortion procedures, which are highly controversial and have serious moral, ethical and social complications. No one, including doctors, have unrestricted autonomy that allows them to act in whichever way they wish.

Occasionally, pro-choicers will bring up the fact that fetal cells often remain in the mother for years after the pregnancy (Dawe et al.). They are trying to obfuscate the issue and discredit the separate nature of the two human beings, mother and child. Their argument goes something like this: Cells with fetal DNA stay in the mother for years after the pregnancy. If one were scientifically able to identify and remove these cells, it is clearly not a problem. In the same way, if one were to identify and remove fetal DNA cells earlier, say just a couple weeks into the pregnancy, then one is still not doing anything wrong or intrinsically different. It doesn't matter when fetal cells are removed, because they are just cells inside of the woman.

The difference is glaringly obvious. The "clump of cells" at the beginning of pregnancy is entirely different from the fetal cells that remain in the mother. The "fetal cells" they are referring to in the early pregnancy is a unique, individual human being and that is his/her body. The "fetal cells" that remain in the mother are not the personhood of her child and are essentially no different from when someone receives cells in a blood donation. Yes, they have the child's DNA, but the cells are not the actual child. The remaining fetal cells could never have even come about if the embryo had not already come into existence. What makes pregnancy so unique is that there is now present a brand new life who just needs time and basic nourishment to grow into a recognizable form.

The medical community clearly acknowledges the difference between these two sets of "fetal cells," defining "pregnancy," "embryo" and "microchimerism" very differently. Microchimerism can occur from pregnancy, transfusion or transplantation and "is the presence of a small population of genetically distinct and separately derived cells within an individual" (Dawe et al.). These fetal microchimerism cells are not human beings at the tiniest stage of development and, thus, do not need to be treated with the same amount of care and respect. In fact, the microchimerism goes both ways with both the mother and the fetus "exchang[ing] cells and genetic material such as DNA" (Comitre-Mariano et al.). This in no way changes that the mother is her own human being with a unique genetic code, and the fetus is his/her own human being also with a unique genetic code; just as a blood transfusion recipient remains separate from the donor. Additionally, this phenomenon of fetal microchimerism actually "may have a beneficial (protective and regenerative) role in maternal health" (Cómitre-Mariano et al.). While the mother gave so much of her body for the health and life of her child, the child's cells can continue to provide a positive impact on the health of the mother for a long time after birth. What a beautiful example of God's glorious design!

11.

Career

Common Ground Possibilities:

- We both agree that women should not have to choose either motherhood or a career.
    - Their career and dreams should not have to be sacrificed and do not need to be mutually exclusive of motherhood.
- We both agree that companies should offer paid maternity leave.
- We both agree that companies should be accommodating and offer working mothers flexible hours, particularly now that so many people work remotely anyway.
- We both agree that there should be affordable childcare options.

Response:

It is wonderful when women achieve success in the workplace. It is also wonderful, in a very different way, when a woman becomes a mother. Both are great occurrences and the, literally, fatal mistake happens when women are lied to and told that they are mutually exclusive. The solution to the difficulties of balancing motherhood with one's career is NOT to take away one's motherhood. Actually, an abortion cannot take away one's motherhood at all. She will be a mother for all of eternity. Abortion kills the child and re-moves the reminder, but that woman will always be a mother no matter what. The pro-life position holds the key to the real solution. We support real changes in the community to make motherhood more manageable for women who choose or have to work. (This is where common ground really comes into play.) And, even if the woman is in a situation where giving birth to the child will slow her career; though unfortunate, it is not justification enough to end the life of that child. Just because someone is making achieving your goals harder, does not give you the right to kill him/her. Life throws many twists and turns in everyone's paths, and the way in which one responds highlights his/her true character. The idea that women can only achieve success in their careers if they kill off their children, is the farthest thing from true feminism. It is truly absurd. Women should be insulted when told that they need abortion for their ca-reers, while men are becoming both fathers and business men on a daily basis. If men are not expected to choose between successful ca-reers and their children, why should women? Imagine if society told men that they needed to go through intense medical procedures and kill their offspring or else they would never get promoted. As out-landish as this thought is for men, it should be equally unthinkable for women.

12.

Education

Common Ground Possibilities:

- Title IX should be more closely followed so as to better support pregnant and postpartum women (Title IX).
- Schools should enact policies to help mothers after the childbirth stage, so they can continue to go to school while parenting.
- It would be wonderful to see an increase in college scholarships for pregnant and parenting students.
- Schools should have the option of offering a slower timeline to obtain one's degree, so that one can also fulfill his/her parenting responsibilities.

Response:

Our society should not be putting young women in the position of continuing her education or continuing her pregnancy. There is absolutely no reason that parenting and graduating should be mutually exclusive. Both motherhood and education better society as a whole and develop one's maturity. Our communities improve when mothers are well educated, critical thinkers, who can teach their children multiple valuable life lessons. Likewise, our society is better off with mothers, children and families. When well educated people have families, their families bring a unique perspective to various topics, not to mention the joy and sense of community that families foster.

That being said, pregnant and parenting students face a lot of difficult challenges in obtaining their degrees. The best response that our communities can give these women is to validate her educational goals and motherhood simultaneously, and help her with the practical steps to do both well. This is truly the pro-life and pro-woman response. Pro-choicers pressuring the young girl to get an abortion and scaring her with the idea that she will never graduate, could not be further from being pro-woman. This sets the stage with a false narrative. Just because something is difficult, does not mean that it is impossible or not worthwhile. The unborn child's life has worth. The girl's motherhood has worth. The girl's education has worth. When all these factors come together in unison, then everyone thrives. (In a debate, this might shift the conversation back to a question about when human life begins.)

13.

Finances

Common Ground Possibilities:

- We both agree that situations can be really difficult and financial hardships can bring about a ton of stress and affect lots of areas of life.
- We both support community agencies that help alleviate financial hardships.
- Pregnancy Medicaid is a great resource for women, since medical bills are one of the biggest expense categories.
- Helping women in need enroll in WIC is a wonderful way to alleviate some of the grocery and formula bills.
- Paid maternity leave should become common practice.

Response:

A survey by the pro-choice Guttmacher Institute, found that 73% of women listed "can't afford a baby now" as one of their reasons for getting an abortion (Finer et al. 113). Such a large percentage of women all having the same problem should make us pause and think about how the pro-life position can offer some solutions to this large obstacle. Many pro-choicers have genuinely caring hearts and are not going to be willing to join the pro-life movement if we do not likewise show genuine care and concern for those women going through tough times. By offering some common ground and solutions for providing tangible assistance to these women, we are not only showing the pro-choicers that we are truly pro-woman, but also we are doing the right thing. The pro-life movement truly is pro-life; pro the life of the baby and simultaneously pro the life of the mother.

Building on this model of valuing the life and the quality of life of the mother, the pro-life movement seeks to directly alleviate the mother's financial struggles instead of providing the supposedly "quick fix" of abortion, which does not actually do anything to help her difficult situation. A woman in financial hardship who has an abortion, is still going to be a woman in financial hardship. The root cause of her problem will remain. By helping an expectant mother get the resources that she needs, the pro-life movement is acting more authentically pro-woman and is truly caring for her and her needs. Practical ways by which pro-lifers can do this is by getting her in contact with community organizations that specialize in assisting people with her specific needs. Different organizations have different specialties, but many communities have ways in which to help with food, rent, utilities bills, childcare, clothing and such. This solution actually solves, or at least begins to solve, the underlying issue.

Not only does abortion fail to mend the original problem, but also it heaps a new financial burden on the women. Abortion is expensive. Depending on the gestational age and type of procedure, it can cost hundreds to thousands of dollars. Added on to this base cost can be expenses such as gas, plane ticket, food, hotel (and other out of state costs), child care, medical complications cost, and later therapy; not to mention, the physical, emotional, psychological, and spiritual costs. Clearly, this is by no means the easy and affordable option that the pro-choice side makes it out to be.

Quite bluntly, the pro-choice side is discriminating against low-income women. Instead of putting in extra effort to make these women's lives better, they lie to them saying that their best option is to abort their children. These lies come from the abortion clinics, friends, and even the father of the baby. In a survey of 1,000 men, 14% admitted that they encouraged their woman to get an abortion because they "didn't want to pay child support" (Care Net 7). Making a woman feel like abortion is the only feasible financial option is the antithesis of what it means to be pro-woman. It forces her into a demeaning position where she is pressured into choosing between the life of her child and providing for herself. This should never be an either or situation. It is unjust for a woman to have to make such a large life decision, that will affect every aspect of her life - emotionally, physically, familially, and spiritually; merely on the basis of whether or not her bank account will balance.

Sometimes it is easier to highlight the seriousness of the situation by comparing the pregnant woman to the mother of a toddler. Everyone should find common ground with the idea that a mother could never kill her toddler in order to alleviate the family's financial stress. Obviously, this would be absurd and horrific. This horrificness is pertinent no matter the age of the child, 5 years, 4 months, 3 weeks, 2 days, or 1hr. This horrificness is also present if the child is in utero. Even though they have not been born yet, they are still a member of the family. Why should the youngest member of the family have to die because serious financial stressors are present. (In a debate, the conversation will probably then shift back into a discussion about when life begins.)

Just as a woman should not feel pressured into abortion for financial reasons, she also should not feel pressured into adoption solely because of monetary concerns. That being said, adoption can be mutually beneficial in many situations, as it allows the child to live and be raised in a loving home, while freeing the birth mom from the expenses of the pregnancy, birth and cost of raising a child.

14.

Other Children

Common Ground Possibilities:

- Having multiple young children can be challenging and no one should have to do it without a support system.
- It's hard when your plans change and it feels like your life is getting turned upside down.

Response:

29% of women wanting an abortion cited "the need to focus on other children" (Biggs et al.). Clearly, this is a very prominent recurring theme. The most direct response would be to point out that a woman does not have the right to end the life of another human being even if the human being is an inconvenience and would make her feel overwhelmed. However, phrasing your response like that would lack compassion and charity. Truth and charity always need to go hand in hand. Remembering this, it is helpful to focus on the root cause of her stress, just as we did with concerns about school, career, and finances. An abortion does not take away any of her current stress nor do anything to calm her life down. If she has an abortion, she still has to return to her hectic and crazy life. Not to mention the additional hassle of going to the appointments and dealing with the physical, hormonal and emotional side effects of the abortion. Her worries about caring for her other children are understandable and actually highlight how much she cares about them and how seriously she takes her role as their mother. Her current strengths in motherhood should be reaffirmed and the comparison can be drawn between her mothering of her older children and her already present motherhood of this youngest child. When she is supported in her motherhood role, her confidence grows and she is better equipped to put her whole heart into caring for another new life.

There are a few different ways that concerns over other children can manifest themselves. If the claim is that the woman is doing okay with her current kids, but that one more would tip her over the edge - then that is a fear based hypothetical argument. We have no way of knowing what our future holds or how someone will affect our life. We can only guess. Good decisions are made when one can calmly and rationally think about an issue, not when one is crippled in anxiety and fear of the unknown. A choice made in paralyzing fear of the future is no real choice at all. A real alternative would be to help and support her in various ways through the next few years, so that she is not alone.

Gently, putting the timeframe of childhood in perspective can also be very helpful. While having multiple young children, who all need a lot of love, time and attention, can make for very long and difficult days; it is a phase and not a long lasting issue. Within a few months the baby usually learns how to sleep for a respectable stretch of time and then shortly thereafter learns how to feed himself/herself. They quickly learn how to entertain themselves and their siblings, lessening the need for mom to be the playmate. All this to say, they do not stay in the high-demand infant phase for the 18 years until they go off to college. They become more self-sufficient and grow to rely on their siblings as well.

Sometimes women do not want to admit that they would be overwhelmed with the growing number of children, so they shift the focus to their older children and claim that it is in their best interest for her to abort the current pregnancy. This way she can spend more of her time and energy on them. This is another falsehood, rooted in fear. Oftentimes, siblings grow up being each other's best friends and testify how they could not imagine life without each other. While adding a new member to the family takes time and adjustment for everyone, everyone is capable of making that adjustment and many unexpected joys and blessings will be uncovered in the upcoming years. Again, it is important to remember that the question at hand is not whether or not to add a new member to the family, the question is what is going to happen to this already present youngest member of the family. The youngest child should not have to die in order for the older siblings or parents to feel less overwhelmed. Like we have discussed previously, a toddler would never die in order to alleviate family stress. The toddler is beloved by the family, has memories with them and is visibly human. The unborn child is fully human and will become more visibly recognizable with time. The memories and feelings of love will also develop with time.

The most important thing to remember is that one never knows what the future is going to hold. Fears and problems that are very real in the current moment, might dissipate quickly. Do not prematurely assume that the worst case scenario is going to become reality. The unborn child can potentially bring unforetold joy and healing to the family unit. One simply does not know the wide variety of struggles, joys and graces that are awaiting us in the future years.

15.

Rape / Abusive Partner / Incest

Common Ground Possibilities:

- No one should stay in an abusive relationship.
- A horrible injustice was done to the woman.
- She should be given an abundance of support and help with healing.
- She should receive physical and emotional care.
- The rapist should be punished and held accountable to the law.

Response:

The most important thing to remember when discussing rape, is to begin with common ground, such as those listed above. Only after one has restated the above points, is one ready to get into the specific arguments. I find the easiest segway to be talking about how both lives should be valued. Yes, we have common ground in valuing the mother's life, but the pro-life side is simultaneously valuing the life of the unborn child as well. Whether she was raped or the victim of other abuse, the mother absolutely deserves support in the healing process and deserves to feel valued and loved by others. At the same, the newly created life also needs to be respected. People come into the world in all sorts of unfortunate situations, but one's conception does not predetermine one's worth and value. The unborn child has not done anything wrong and is, actually, also victimized by the rapist's actions.

Because the rapist is the one who committed the horrific crime, he should be the one to receive the punishment. This is a great time to reinforce the common ground of punishing the rapist and holding him accountable to the law. The culprit, the rapist, should be the one to pay for the consequences of his actions, not an innocent third party, the unborn. In sentencing criminals for other crimes, the criminal, himself, is left to deal with the harshest punishment and cannot pass that off to anyone else. However, in the case of abortion and rape, it is the child who is quite literally sentenced to the death penalty and thus receives the ultimate punishment. If a baby's father stole an item, was caught, and then was sentenced to jail; no one would approve of the baby taking the father's place in fulfilling the punishment. Receiving the death penalty is even more extreme than fulfilling someone else's time in jail, so why is the unborn child being punished so severely? (Note: This is definitely not a recommendation that the rapist's punishment should be the death penalty, but rather an explanation showing the severity of the consequence for the unborn.) Occasionally, pro-choicers will try to corner you and ask what you think the punishment for rapists should be and then they try to argue with that. This can be simply avoided by saying that it would be up to the legal system to evaluate the circumstances and determine the appropriate punishment. Even in cases where the rapist is held accountable to the law and the woman still decides to abort, the innocent child is still paying the ultimate price for a crime (s)he did not commit.

While the rapist or abuser being punished is a good and necessary thing that can also help with emotional healing and closure for the woman, having an abortion is not going to provide more healing or closure for her. An abortion is not a magical solution that can unrape a woman. It cannot not fix the original problem. Nothing can be done to turn back the clock and change history. The only thing that can be controlled now is her response in the moment and the trajectory she chooses for her future. Not only does an abortion not help the woman heal, but it can inflict more trauma on her. Despite the pro-choice side's insistence that abortions are quick and easy, they actually take a large toll on the woman's body. Even the pill form of abortion has a large effect on the woman, because expelling an embryo is an involved and unnatural process for her body. Further, if she chooses a procedural abortion, then she is subjected to a very invasive procedure. This procedure is very uncomfortable and vulnerable for many women; but how much more so would it be after having just been raped, where your privacy and autonomy were completely invaded and disrespected. Assumedly, this could easily trigger flashbacks and panic. Not to mention, abortions have several very common side effects that are undesirable at any time, most of all right after an act of rape. Her body is expected to go through another physically intense, often very painful, process. Further, any pregnancy is a deeply personal event and the abrupt ending of one (even an undesired one) is bound to have an emotional impact on the woman. Complex feeling and emotions would only be compounded when one is simultaneously processing and recovering from rape and abuse. An abortion inflicts more pain and suffering on an already vulnerable woman. It adds another long lasting trauma to her life just days after she has already endured unimaginable trauma. Simply put, two wrongs, do not make things right.

While the statistics of rape should not be the first point you bring up, because you do not want to discredit their concerns, sometimes it is helpful to put the numbers in perspective. According to the pro-choice Guttmacher Institute, cases of rape made up 1% of abortions and cases of incest were .5% (Finer et al. 113). This leads to the question, "Why should the United States keep abortion legal for everyone, if it is such a small minority that have the extenuating circumstances of rape and incest?" Try asking the other person if they are okay with the rest of abortions being banned, if exceptions were made for rape and incest. If they say yes, then some common ground (the idea that the rest of abortions are wrong and should be made illegal) has been found. Yay! Next, the above explanations may need to be reiterated in order to help them become pro-life in these situations, or the conversation might be over for the day and maybe the next conversation can build on that foundation of common ground. If they say no, that actually abortion should not be banned no matter what, than the conversation does not need to be focused only around rape and incest and will instead shift to other, broader talking points. Ultimately, abortion should not be nationally legal, because 1.5% of abortion minded woman are struggling with the pains of rape and incest.

Based on the feel of the current conversation, sometimes it can be helpful to bring up a counter narrative. Instead of the baby being thought of as the "rapist's baby," one can gently point out that the baby is very much "her baby." Even though the father committed a horrible crime and his identity might even be unknown, the identity of the mother is known and she is one incredible and amazingly strong woman. Any child would be proud to call a woman of such resilience and strength "mom." Nothing, not even an abortion, can make her no longer a mom. She already is the mother and the baby already is developing a loving bond with her. In utero, the child is unaware of the rape, but is very aware of his/her mother and is deeply connected to her. The woman should be given extra encouragement and support in her new journey of motherhood. It is actually a very belittling and unfair presumption to assert that the woman would not be strong enough and capable of handling the pregnancy and, if she chooses, parenting. The woman has already shown incredible strength while enduring so much hardship and suffering. Her family, support system, and community should work to build up her strength and self-confidence, not instill doubt about her abilities.

That being said, the woman could very easily come to the conclusion that raising the child herself would feel too hard for her or would be too much of a painful reminder. This would be entirely understandable and is where adoption can be a really beautiful solution. Adoption can simultaneously love and celebrate the newly created life, while also respecting the woman and helping her heal and find closure after the rape. She can have the distances that she desires and needs, but the child's life is still respected and has a chance to blossom. When parenting or choosing adoption, the woman should still receive an outpouring of love and support.

On rare occasions, the pro-choice side will assert that it is better for the mother AND the child to have an abortion. They claim that the child's life will be too tainted by the event to be happy or that the adoption will be doomed. The absurdity of the claim usually discredits itself. Typically, if you rephrase the claim back at someone and ask if this is truly their position, they will realize how extreme that claim is. However, if they stand by the claim you can point out: The above discussion points about the intrinsic value of one's life. The beauty of all life, even one with challenges. All life is worth living. Everyone has a basic right to life. It's a highly subjective claim (at best) and the unborn child could very easily disagree with the claim and very much rather be given the chance to live. Society does not go around killing other, born, people because someone else determined them to be better off dead. Even if someone desires to be dead, a good society helps the person through their current problems and does not aid or speed up their suicide, let alone assist someone else in their murder. Overall, the claim that abortion is in the best interest of the child, is farfetched, extremely subjective and devoid of ethics and morality.

Lastly, when discussing post-rape care for women, it is crucially important to note what time of her cycle she is in. There is a fair chance that she is only approaching her fertile stage and has not ovulated yet. In these cases, the United States Conference of Catholic Bishops has put out a directive clarifying how best to treat a post-rape woman.

> "A female who has been raped should be able to defend herself against a potential conception from the sexual assault. If, after appropriate testing, there is no evidence that conception has occurred already, she may be treated with medications that would prevent ovulation, sperm capacitation, or fertilization. It is not permissible, however, to initiate or to recommend treatments that have as their purpose or direct effect the removal, destruction, or interference with the implantation of a fertilized ovum" (Bransfield 15).

This key distinction is imperative because it distinguishes between preventing conception, which is permissible in cases of rape, and preventing implantation or the continuation of an already present pregnancy, which would be an early term abortion. In this way, the USCCB seeks to honor the dignity of the woman and help her heal, while also respecting any new life that has already come into existence. Overall, this idea sums up the pro-life position in response to rape: respect both lives and seek to help both have the brightest future possible.

16.

Medical Condition of Baby

Common Ground Opportunities:

- Receiving a fetal medical diagnosis is a really hard thing for the parents, and for everyone involved. They should be given extra support now and in the future.
- Companies should work with their employees to provide good medical insurance to help defray the additional expenses that the family would be incurring.
- Communities should work to develop programs that assist those living with disabilities and their families.
- Businesses should create affordable technology to aid caring for these individuals in their homes and to ensure the highest quality of life.

Response:

Conversations about this topic are typically fraught with high emotions that, unfortunately, cloud people's ability to view the circumstance from an objective ethical and moral perspective. It is the tricky role of the pro-lifer to be compassionate and understanding of the difficulties surrounding the situation, while also upholding the dignity of the unborn human life. These two always need to go hand in hand. Yes, some new life brings large challenges for a family. However, one's worth, and thereby right to live, is not determined by whether or not your life is filled with obstacles or brings challenges to other people. Each one of us has intrinsic, God-given worth that no one can take away and nothing can diminish.

Margaret Sanger notoriously wrote in her 1921 *Birth Control Review*, "The most urgent problem today is how to limit and discourage the over-fertility of the mentally and physically defective" (Sanger). What she meant by "defective" was cloudy, but what was abundantly clear was her assertion that some people had more worth than others and that those deemed less worthy should be deprived of their right to procreate. This is inexcusably an extreme case of discrimination, judging one's worth by one's health or another's preconceived notion of your health. This hurtful mentality is still grossly prevalent in today's abortion machine. In a 2009 interview with the *New York Times*, Justice Ruth Bader Ginsburg admitted, "Frankly I had thought that at the time Roe was decided, there was concern about population growth and particularly growth in populations that we don't want to have too many of" (Bazelon). Like Sanger, her idea of undesirable populations was extremely vague and worrisome. These women leaders of the pro-choice movement are not alone in their radical ideology. The Joint Economic Committee of Republicans Report of 2022 found that selective abortions on babies with a sickly prenatal diagnosis was radically reducing diversity. They estimated that 217,000 people with Down Syndrome would be missing in the United States over the course of the next 50 years (Lee Down Syndrome 1). Equating health and worth, and having such overtly discriminatory practices are two large blotches on the abortion industry's record.

One's health or physical abilities cannot determine one's worth. For if it did, then people who are even temporarily sick would be viewed as lesser and, consequently, would be in danger of losing their lives. Imagine the public outrage that would rightfully occur if a gunman made his way into the ICU unit of a hospital and started shooting everyone. If his defense was, "They were just going to die anyway, so I put them out of their misery faster and freed their families from ongoing financial hardships," no one would view that argument as valid and a proper justification. Everyone would be devastated and grieving the loss of life. When someone is sick, they deserve extra love, care and medical treatment. Even if someone is almost certainly going to die from a disease and natural causes, we do not intentionally end their life earlier. Similarly, the unborn deserve to not have their life abruptly ended by someone else who views their pain and suffering as too much or too expensive. Even a very serious, life long and life threatening condition does not diminish the inherent value of a person's life.

Occasionally, some pro-choicers bring up and support physician assisted suicide and use it as a justification for abortions related to medical issues of the baby. If they are going to that level, that might signal the natural end of the conversation; as it would be hard to have fruitful dialogue when someone's beliefs are on the far extreme end and set in stone. However, while in no way agreeing with physician assisted suicide, you could try pointing out that with an abortion, the person who has the medical condition and whose life would be ending early, is not the one who is making the decision. The unborn person has no voice and is being denied the chance to experience some of the joys of life even for a short time.

The idea that health equates value and, therefore, abortions are permissible if the unborn child has a disease, is a very ill defined slippery slope. Where does one draw the line? Life ending illness? Life limiting illness? Down Syndrome? Missing a limb? Male instead of desired female (or vice versa)? In the continuation of this thought process, anyone with anything deemed any level of debilitating or not desirable could be aborted.

Many people have already gone down this slippery slope and choose to abort based on the sex of the baby. This problem is prevalent enough that, as of August 2023, 11 states had enacted a ban on sex selective abortions (Abortion Bans). If abortions because of sex selection never happened, then no state would have to ban it. But, if they happen enough that some states have banned it, then why are the majority of states not also creating bans? (Note: Many states are rapidly changing their laws in the post-*Roe* era.) Unfortunately, the exact stats for sex selection abortions in the U.S. are not available because "Gender is not recorded in the two primary abortion surveys conducted in the United States, conducted by the Centers for Disease Control and Prevention and the Alan Guttmacher Institute" (Abrevaya 8). They are intentionally choosing to turn a blind eye to this discriminatory practice.

Sex selective abortions are also prevalent globally. So much so, that The World Health Organization felt the need to come out with an entire brochure titled, "Preventing gender-biased sex selection." In this brochure, they state that "Sex selection… typically occurs because of a systematic preference for boys" (Preventing Bias V). "The biologically normal sex ratio at birth ranges from 102 to 106 males per 100 females. However, ratios higher than normal – sometimes as high as 130 – have been observed" (Preventing Bias V). An article in the 2015 *Population and Development Review* concurred that the world is missing many women and calculated that, "The global annual number of missing female births rose from near zero in the late

1970s to more than one million per year in the period after 1990, reaching 1.6 million per year in 2005–2010… The cumulative number of females missing at birth between 1980 and 2010 exceeds 30 million" (Bongaarts and Guilmoto 253). Now in 2025, in China alone, there are approximately 35 million more males than females (What happens when 30). This leads to countless problems, including concerns about finding partnership and marriage, future birth rates, and a generally well functioning society. Worldwide, whether this is from sex-selective abortions or also sex-selective IVF, which likewise causes the death of countless embryos, this is the exact opposite of the "feminism" which the pro-choice side so loudly heralds.

More widely discussed are those abortions which occur because of a life limiting illness, such as Down Syndrome. Down Syndrome frequently makes the news as one of the few diseases which have been nearly or completely "eradicated" in several European countries. How is it that no one in these countries is being born with this disease? Heartbreakingly, it is not because a phenomenal new cure was found, but rather because the pregnant mothers choose to abort after receiving a prenatal diagnosis. For example, in Iceland, "nearly 100 percent of pregnancies that receive a positive test for Down syndrome are aborted" (Lee Down Syndrome 2) In the U.S., the termination rate is between 61% to 93%, depending on the study (Natoli et al. 2011) While lower than Iceland, it is still appallingly high. While individuals with Down Syndrome have unique challenges that most of us never have to face, that in no way lessens the beauty of their lives. The Joint Economic Committee of Republicans noted that 99% of these individuals "indicate that they are happy with their lives" and that 79% of people who have a relationship with them have a more positive outlook on life because of them (Lee Down

Syndrome 6). If you have ever been around one of these remarkable people, you have undoubtedly noticed their overflowing joy and kindness of heart. Their uniqueness blesses this world in countless ways.

Aside from the fact that those with certain diseases and conditions still have inherently valuable lives, some of the prenatal screening tests have questionable accuracy. Genetic non-invasive prenatal screening tests work by drawing blood from the mother and then analyzing the fetal DNA. The results that the woman receives tell her if the unborn child is more at risk for various conditions. In other words, these tests are only screening tests, not diagnostic ones. The FDA also points out that these tests cannot distinguish between an abnormality in the fetus and one in the placenta (Non-Invasive). This is of the utmost importance because when a woman chooses whether or not to have an abortion based on this information she is acting on very limited information and is playing the probability. In 2022, the FDA came out with a strong statement warning women and health care professionals about the potential for inaccurate results, highlighting the vast implications of this and reminding everyone that none of these tests had been approved by the FDA (Non-Invasive). While some diseases and conditions do receive an official diagnosis in utero, it is important to distinguish between these cases and the ones from the screening tests, which have become increasingly popular. While these tests are often performed earlier in pregnancy, sometimes the actual abortion is not carried out until the third trimester (Borgatta). At this point, the woman will typically go through the stages of labor and delivery with an induction abortion. The abortionist chooses this method so that the fetal body is more intact, compared to the D&E, and they can confirm the prenatal diagnosis (Shulman et al.). Imagine examining the remains afterwards only to discover that the prenatal testing was incorrect, as the FDA has warned that it might very well be. Even if the diagnosis was confirmed, the direct and intentional ending of the innocent human life was still not justified.

The most heart wrenching cases are those in which the unborn baby has a life-ending disease and will die immediately after birth, or even still in utero. When discussing these cases, we should approach the topic with an abundance of compassion and empathy. It is understandable why some women would wish to have an abortion and speed up a supposedly inevitable outcome. It can be very trying when the process of losing someone is extended over a long period of time. This is where ethics, morality, and natural law really come to the front of the conversation. Just because something may be tempting, does not mean that it is permissible from an ethical, moral and natural law perspective. Having an abortion adds a new layer of heartache, since the woman will now share culpability in the child's death. While situations of death immediately after delivery are tragic and traumatic, that life is still inherently valuable and a God-given blessing. As the Author of all human life, it is God's role to dictate when each one of us is called home to Heaven. When we "speed things up," we usurp God's role. Even a short life is still a beautiful life. Most of us do not know when our lives will end. Just because some people get a warning, does not mean their death should be sped up from the near future to the present. This leads to another very slippery slope. How short of a life expectancy justifies abortion? What if the child is expected to live for 1 hr? 1 day? 1 week? 1 month?

1 year? 10 years? Instead of pressuring women in these heartbreaking situations to have an abortion, families and communities should be coming together to support them, grieve with them and love them. Perinatal hospice and palliative care can provide some beautiful services during this time to help honor the shortly lived life and to help the family cope.

Since medical diagnoses are really difficult for all involved, how can we respond with support and love? There are three responses to keep in mind. First off, the woman and her family will need extra assistance during the pregnancy and delivery. Based on your relationship with them, this could take multiple forms: meals, babysitting, phone call chats, or helping her do research, are all good places to start. On a medical level, she and the baby should be seen by specialists for the specific diagnosis in order to make the most life affirming plan possible. Secondly, if she chooses to parent the baby, herself, then she will need continued tangible help. This could include good insurance coverage to cover the increased medical expenses, access to community programs both for parental encouragement and for child development, and availability of affordable technology to make day to day life easier in the home. Thirdly, if she would prefer to make an adoption plan, she should be told about all of the wonderful adoption agencies that specifically are geared towards finding matches for children with a medical diagnosis. That child, no matter her/his diagnosis, will be very much loved, wanted, and treasured by an adoptive family. In general, our society needs to be more aware of the beauty of adopting these special needs children. They are not a burden, but rather a great blessing to be cherished for eternity. See adoption section for a list of agencies.

Molar Pregnancy:

Molar pregnancy is an extremely rare occurrence that happens "in about one in 1000 pregnancies in the United States" (Paul et al.) There are two types: complete and partial.

Complete molar:

A "complete mole arises when an empty egg is fertilized by 1 or 2 sperm" (Cue et al.) There is no maternal nuclear DNA, only mitochondrial DNA, resulting in a "complete lack of embryonic or fetal tissue" (Cue et al.) Simply put, there is no baby.

However, very rarely there is a "complete hydatidiform mole with a coexisting viable fetus" (Dolapcioglu et al.). "The chances of a live birth have been estimated between 30 and 35%" (Dolapcioglu et al.). As of 2015, of the 200+ documented cases, there have been 56+ records of a live birth (Sargin et al.). Though not a high chance of survival, this percentage is enough to provide hope. Women with these pregnancies should receive expert medical care and extra monitoring because they are prone to more complications, including some very serious ones (Ghalandarpoor-Attar and Ghalandarpoor-Attar). Delivery may have to be induced early in order to best care for and respect both the mom and baby.

One of the most common concerns stemming from a complete molar pregnancy is it turning cancerous. In about 15% of cases with a complete mole, the mom will need to undergo chemotherapy (Paul et al.). These women should be given an abundance of love and support during this very difficult and painful time in their life. It is a very hard thing to undergo cancer treatment at any time, let alone right after you lose what you thought was your baby/pregnancy.

Partial molar:

Thankfully, the chemotherapy rate for pregnancies with a partial mole is much lower at 0.5% (Paul et al.). Overall, there is usually a smaller complication rate with partial molar pregnancies and they are less common (Parial Molar). So what is a partial molar pregnancy?

Typically, "A partial mole arises when a viable ovum is fertilized by 2 or more sperms" (Cue et al.). There is a baby present and a heartbeat can sometimes be detected (Cue et al.). Oftentimes, a miscarriage will happen naturally with "the fetus usually [dying] within a few weeks of conception" (Cavaliere et al.).

("Very rarely, a partial molar pregnancy develops with two maternal and one paternal haploid set... (and) have been reported to result in live births, with subsequent early neonatal death") (Cavaliere et al.).

Management of pregnancy:

Well, what if the pregnancy does not end naturally in a miscarriage after the first couple weeks? A 2021 medical journal confirmed that, "If there is a euploid fetus in a singleton partial molar pregnancy... continuing pregnancy is possible" (Ghalandarpoor-Attar and Ghalandarpoor-Attar). "In some situations molar changes in placenta is associated with a normal diploid fetus" (Rahamni and

Parviz). When this happens, an ultrasound is key in making the diagnosis (Rahamni and Parviz) and then "serial beta HCG monitoring and frequent scans are the tools for management" (Surendran et al.). If there are severe complications for the mom, then an early induction, with expert NICU care given to the preemie, may be done in order to value both lives. In cases where there is no heartbeat and the baby has died, but physically remains inside the mother, then Dilation and Curettage are commonly performed. Sometimes a hysterectomy is also done (Cue et al.) (Oftentimes, pro-choicers will advocate for the D&C procedure regardless of the specifics of the individual pregnancy.)

There have been several documented cases of a live child being born along with a partial mole diagnosis.

- A "normal- appearing male fetus with diploid karyotype was delivered at 31 weeks gestation" and "pathological examination of the placenta showed changes of partial hydatidiform mole" (De Franciscis et al.) Both the baby and mom were found to be healthy at the 12 month check-in (De Franciscis et al.).
- There was a "multicystic" and "partial molar placenta in which a live female baby was delivered at 32 weeks gestation" (Hsieh et al.) She was delivered via c-section and was originally anaemic but recovered after 2 weeks (Hsieh et al.).
- A baby girl was born at 34 weeks GA via emergency c-section. There had not been any complication until delivery, but "molar changes were also confirmed by histologic examination" (Ghalandarpoor-Attar and Ghalandarpoor-Attar). After follow-up, "the patient was completely healthy, there was

no GTN, and her neonate had normal growth and development" (Ghalandarpoor-Attar and Ghalandarpoor-Attar).

Risks to baby:

While there have been a few rare instances of a healthy born baby, partial molar pregnancy "is commonly associated with congenital fetal malformations" (Hemida et al.), aka "severe birth defects" (Surendran et al.). These kinds of life limiting diseases are a very sad and difficult diagnosis for parents to face. Experts in the fields of perinatal hospice and palliative care can help walk alongside the parents during this extremely trying time. A trusted doctor, who will recognize the worth of this tiny life, should be seen so as to find a path forward that is respectful of that new, very vulnerable life. Even a life that is only lived out of utero for a few minutes is still irreplaceably precious. None of us know how long we are going to live. Each day is a gift.

There are several instances of families who have chosen to cherish these little lives for as long as they could, even when faced with extreme diagnoses. Date is in correspondence to the journal source and order is arranged beginning with longest extrauterine life.

- 1988, A Caucasian baby girl was born with tetraploidy and was still alive and "growing well" at age 22 months (Lafer and Neu). The research team wanted to provide hope for other parents and noted that, "When counseling parents (prenatally) it is important to realize that, although quite rare, tetraploid individuals can be born alive and live for at least 22 months" (Lafer and Neu). Follow up unavailable for how long she lived.

- 1986, A baby boy with triploidy continued to live for 10.5 months. Researchers noted that, "improved survival, (could be) possibly due to better management of respiratory illness and prematurity" (Sherard et al.).

- 2005, A baby with triploidy was born and continued to live for 164 days. This baby's survival time was "the fourth longest worldwide" for this condition (Iliopoulos et al.)

- 1977, A baby girl with triploidy was born and continued to live for 160 days (Cassidy et al.). "Previously, the longest survival of a non-mosaic triploid had been 9 days" (Cassidy et al.).

- 1996, A baby with "mosaic trisomy 8 with a tetraploid cell line" was born and continued to live for 14 weeks" (Roberts et al.)

- 1993, "A girl with full triploidy and multiple malformations, (was born and) [she] survived for 10 1/2 weeks," which is much longer than the expected few hours (Niemann-Seyde et al.).

- 1999, A baby girl with triploidy "survived for 46 days" after birth (Hasegawa et al.). As of 1999, "five triploid infants, including the girl we described, who survived 4 weeks or more" (Hasegawa et al.). (Others of which might also be listed individually in this list.)

- 1984, A baby girl with "triploid[y] survived 45 days" after birth (Maraschio et al.).

- 1980, A baby was born prematurely, but alive with complete triploidy (Rico et al.).

-   1983, Two babies were born prematurely, but alive with homogeneous triploidy. They did pass away from the condition. (Gouyon et al.).

Risks to mom:

Since partial molar pregnancy is such an extreme diagnosis for the baby and pregnancy, how does it affect the mother? Thankfully, "the presentation of a partial hydatidiform mole is usually less dramatic than that of a complete mole" and usually presents similarly to a miscarriage (Cue et al.). For example, "persistent trophoblastic disease or malignant complications are much more common with a complete molar pregnancy (8%) than with a partial hydatidiform mole (0.5%)" (Cavaliere et al.). And, "the outcome of a partial hydatidiform mole after uterine evacuation is almost always benign. Persistent disease occurs in 1.2% to 4% of cases" (Cavaliere et al.). That being said some of the maternal risk factors are: "abnormal bleeding, preeclampsia, eclampsia, hyperthyroidism, anemia, persistent gestational trophoblastic disease, preterm delivery and abruption" (Surendran et al.). But again, such "Obstetric complications… are more common in complete mole than partial mole" (Sargin et al.). As with any high-risk pregnancy, the doctor should take extra caution to make sure that everything is progressing in a manner that is safe for both mom and baby, since both lives are so very valuable.

Overdiagnosis:

Cases of a live baby along a partial molar pregnancy are incredibly uncommon and are currently thought to be "occurring in 0.005

to 0.01% of all pregnancies" (De Franciscis et al.). However, the diagnosis of such a condition can be difficult, which can lead to skewed numbers. While "ultrasound typically reveals anechoic cystic clusters for complete moles", it shows a fetus/baby in cases of a partial mole (Cue et al.). Sometimes, "clinical signs are almost completely absent" (De Franciscis et al.). "Diagnosis is confirmed through histopathological examination after tissue removal." (Cue et al.) Confirmed by (Cavaliere et al.). One of the Australian Departments of Health adds, "A molar pregnancy can only be confirmed when the pregnancy tissue is examined under a microscope by a pathologist. This (is) not always possible as tissue is not always sent to a laboratory for testing after a miscarriage or a normal pregnancy, labour and birth" (Molar). So, considering that an official diagnosis happens from a laboratory examination, and that oftentimes placenta or miscarriage tissue is not sent to a lab unless there is a suspected problem; it seems possible that the numbers of babies born alongside a partial mole diagnosis might be underreported. It seems as if the doctors go looking for a diagnosis after something goes wrong, but do not do the same testing on apparently healthy pregnancies. This potentially allows for cases with a happy ending to be underreported.

Additional Notes:

- "Exposures which independently and significantly predicted increased risk for partial molar pregnancy included... oral contraceptive use for > 4 years" (Berkowitz et al.).
- Complete or partial molar pregnancies should not be confused with the term "chemical pregnancy" which refers to a

miscarriage which happens soon after implantation. Often the woman will have a positive pregnancy test, but then no embryo/baby is seen on the ultrasound. It is thought that between "50-75% of all miscarriages are considered to be chemical pregnancies" (Dumitrascu et al.).

17.

Medical / Life of the Mother

Common Ground:

- These cases are utterly heartbreaking and our thoughts and prayers are with the family.
    - Extra support (emotional, financial etc.) for them during this time.
- We want to do everything within our power to save both lives.
- In cases where the mother is sick but not fatally sick, we want to find medical treatments to both alleviate her symptoms and cure the underlying issue.
- Increased coverage for medical insurance so the woman and her family can afford the best treatment.
- Greater access to specialists for specialized care.
- Screenings available to find conditions earlier.

Response:

As in all of these really tricky cases, the best way to start is common ground. Yes, we care about the baby's life, but we absolutely care about the mother's life and overall health as well. Our expectation of medical professionals is that they will care for both the mother and her unborn child as patients who deserve top quality care. Unfortunately, this is not everyone's goal and the Institute of

Medicine has attempted to vaguely redefine health care as "the degree to which health services for individuals and populations increase the likelihood of desired health outcomes and are consistent with current professional knowledge" (Jones and Weitz). Such wishy washy definitions are the gateway for justifying abortions and considering a wide spectrum of factors as "health concerns".

When discussing cases of health and life of the mother, perhaps the most important distinction is the severity of the condition. There is a wide spectrum of health related reasons that women choose to abort. These reasons range from unpleasant morning sickness to late stage cancer. Because these conditions are of drastically different levels of severity, they should be discussed separately and not equated. Let's begin with morning sickness. Surprisingly, abortions because of first trimester morning sickness are not as rare as you might suspect. A simple Google search reveals that it is actually quite common. Articles such as "Abortion: Last Resort For Hellish Morning Sickness" and "Pregnancy made me so sick that I begged for an abortion" flood the search results (Abortion Sickness) (Howden). As all mothers know, morning sickness can be a real doozy. However, the temporary pain of morning sickness is not justification equal to ending another human life. Rather, the woman's doctor should be working with her to find the right medicine, supplement, routine or whatever else she might need to minimize her symptoms as much as possible. This is caring for the woman's quality of life and health, while still respecting the unborn life. Remember, in order to count as a medical necessity procedure, "The service is not primarily for the convenience of the individual, individual's health-care provider, or other health-care providers" (Studnicki Late-term).

Since health can be so widely defined, many women try to qualify for abortion because of emotional and mental health. This has been happening for a very long time. So much so, that a 1971 study in *The Western Journal of Medicine* found that 97% of women choose abortion for "mental health" (Brenner et al. 22). These terms of mental and emotional health are extremely vague and, thus, can be used to justify nearly anything. In regards to people claiming mental necessity, it is often forgotten that, "There are specific clinical criteria available for determining the medical necessity for psychiatric treatment: a diagnosed disorder... too often, these assessments are neglected or superficially completed using inappropriate documentation and by persons without appropriate credentials and experience" (Studnicki Late-term). Of course, the pro-life side does not want women to be in emotional distress or commit suicide! (This can be another opportunity to reiterate common ground.) However, claiming any level of additional stress and, thereby, a strain on one's mental health, is again not justification equal to ending another human life. This is why some states are adding specific wording to their laws distinguishing between health cases that truly threaten the mother's life and one's that are theoretical (Tanne) (Wehrman).

Next to be considered are diagnosable health conditions that are manageable, and oftentimes temporary, such as gestational diabetes or anemia. Women with these conditions are in need of more medical care, but they are (typically) able to carry their pregnancies to term without seriously jeopardizing their health. Next would be more serious conditions, like preeclampsia. These types of conditions can be safely handled by doctors, but may need urgent care.

Lastly, to be considered are the very serious diagnosis where immediate, drastic action is needed to save the mother's life. Data analyst James Studnicki, who has his Doctor of Science and Master of Public Health degrees, agrees and additionally points out that, "Varying definitions of medical necessity for abortion have ricocheted along a continuum with consideration of a 'broad range of physical, emotional, psychological, demographic, and familial factors relevant to a woman's well being' at one extreme and 'conditions which place a woman in danger of death' at the other" (Studnicki Late-term).

In the extreme and tragic cases where the mother is at risk of dying, then a very important distinction needs to be made. There is a world of difference between an abortion, which directly and intentionally ends the life of the unborn child, and treatment to save the life of the mother, even if the child dies as an undesired result. Neonatologist Kendra Kolb points out that, "A mother's life is always of paramount importance, but abortion is never medically necessary to protect her life or health" (Kolb). Dr. Don Sloan, an abortionist who has performed more than 20,000 procedures, goes a step further and admits to the possibility of an abortion further endangering a woman's health. "If a woman with a serious illness... gets pregnant, the abortion procedure may be as dangerous for her as going through pregnancy.... The idea of abortion to save the mothers' life is something that people cling to because it sounds noble and pure-but medically speaking, it probably doesn't exist. It's a real stretch of our thinking" (Sloan). (Potential maternal risks and complications accompanying abortion procedures are detailed in the above abortion procedures chapter.) Taking into account the views and knowledge of many experienced doctors, the American Association of Pro-Life Obstetricians and Gynecologists firmly asserts that "abortion-removing the fetus with the intent of ending its life-is *never* medically necessary'" (Brief 2021 2). Studnicki elaborates on what counts as "medically necessary" and how pregnancy differs from a life threatening disease. In order to be medically necessary,

> "the service must be required to prevent, diagnose, or treat an illness, injury, or disease. Pregnancy is neither an illness nor a disease and, following conception, is no longer preventable. Therefore, the treatment (abortion) must target another specified illness, injury, or disease. The service must be clinically appropriate and considered effective for the individual illness, injury, or disease. This requirement implies that credible, evidence-based peer-reviewed literature exists that the abortion procedure will produce a positive result on specified outcomes related to the pregnant woman's illness, injury, or disease" (Studnicki Late-term).

From a Catholic perspective, the USCCB has elaborated quite a bit on this stance. In their Ethical and Religious Directives for Catholic Health Care Workers, they concur, "Operations, treatments, and medications that have as their direct purpose the cure of a proportionately serious pathological condition of a pregnant woman are permitted when they cannot be safely postponed until the unborn child is viable, even if they will result in the death of the unborn child" (Bransfield 19). The key distinctions are the "directness," the "seriousness," and the urgent timing. All of these conditions must be met. To further clarify, the USCCB extrapolates, "Abortion, a direct and intentional attack against the child's life, is never morally licit. The unborn child and his mother have equal human dignity and possess the same right to life. When a medical crisis arises during pregnancy, there are always two patients involved. Doctors should do whatever they can to save both their lives, never directly attacking one —through drugs, surgery or other means—to save the other" (Life Matters). Overall, there are two lives that need to be treasured, nurtured, protected and given appropriate health care. Directly and intentionally terminating one's life is never the direct cure for the other.

With this understanding, it can be helpful to look at scenarios trimester by trimester. Perhaps the most widely known and referenced case of life of the mother in the first trimester is ectopic pregnancy. Cleveland Clinic's definition of an ectopic pregnancy is "when a fertilized egg implants outside of the uterus, most commonly in the fallopian tube" (Ectopic Pregnancy Cleveland). This is an extremely serious condition that leads to maternal and fetal death. Thereby, immediate action is necessary in order to save the life of the mother (Ectopic Pregnancy Cleveland). (Although ectopic implantation occurs in only 2% of pregnancies, it accounts for 13% of maternal deaths (Reardon et al.).) Sadly, there is currently no known way to save the unborn life as well. Since there is no chance for the baby to survive and since lack of treatment leads to maternal death, pro-choicers will use this scenario to justify legalized abortion. However, they are leaving out a couple essential facts. Firstly, it is usually the case that the embryo has already died (Condic and Harrison). The embryo almost always dies early on in the first trimester from "insufficient hormone and nutrition supply" (Santiago-Munoz). This alone makes a world of difference because then the doctor would not be directly or intentionally ending the fetal life. In cases where it can be determined that the embryo is still alive, a medical procedure, called a salpingectomy, is available. A salpingectomy is a surgery that removes both the fallopian tube and the ectopic pregnancy inside of it (Ectopic Pregnancy Mayo). Once detached from the mother, the embryo will inevitably die. (In cases, where the pregnancy is implanted somewhere other than the fallopian tubes, ask a trusted pro-life doctor about treatment options) (Huanxiao et al.). Though this is a very sad scenario, it is intrinsically different from an abortion in both directness and intentionality. The embryo is not actively being killed by a drug, forceps, or anything of the sort. Rather, the embryo is left within his/her immediate environment, though that environment is disconnected from the larger support network. Secondly, the intention is not to cause the fetal death but to save the life of the mother, a very noble intention. If it were possible to simultaneously save the embryo, then medical professionals absolutely should. However, with the limits of current medical technology there is no way to save the unborn life, even if left within the mother's fallopian tube. The only potential saving can be for the life of the mother.

The common method for determining if a pregnancy is ectopic or intrauterine is ultrasound. A simple scan can confirm that the pregnancy is developing in the correct location or not. Ultrasounds provide a unique look at that unborn human life that is otherwise hidden. Ultrasounds consistently enable mothers to instantly bond with their unborn children, oftentimes causing her to decide against abortion. When the woman decides against abortion, the abortionist can no longer make money off of her situation. Therefore, many abortionists and clinics have nefarious, financial, ulterior motives for encouraging women to abort. With this vested interest in preventing women from seeing and bonding with the ultrasound, the abortionists discourage women from receiving this basic form of pre-abortion evaluation. In justification, they make the outrageous claim that, "Ectopic pregnancies are uncommon among women presenting for abortion" (Duncan). As if, the pregnancy could tell whether or not it was desired and, therefore, where it should implant! While "approximately 11 per 1,000 pregnancies are ectopic", a Planned Parenthood study found only "seven ectopics per 100,000 pregnancies" in women desiring abortion (Duncan). The conflict of interest in this study is astounding and completely discredits the validity of their claims. Planned Parenthood and the wider abortion industry lies, twists facts, and distorts truth time and time again in pursuit of financial gain. The *Nordic Federation of Societies of Obstetrics and Gynecology* published research calling for, "Initiation of abortion before there is definitive ultrasound evidence of an intrauterine pregnancy in women without signs or symptoms of an ectopic pregnancy should be considered" (Schmidt-Hansen et al.). Similarly, the *TEACH* abortion manual claims that, "If a pregnancy is undesired, there is no reason to delay uterine aspiration to wait for diagnosis; and a diagnostic aspiration will assist in the evaluation of a possible ectopic pregnancy" (Hufbauer et al. 37). An aspiration procedure vacuums out the contents of the uterus, but if there is nothing in the uterus, as in the case with ectopic pregnancies, then the procedure will not complete its goal. Technically, the aspiration procedure might help in the "evaluation" of an ectopic pregnancy by possibly showing that no fetal remains were removed, and thereby they must still be inside the woman. However, this is a very complicated, expensive and roundabout way of doing something that a simple ultrasound could have diagnosed more thoroughly, quickly, and inexpensively. But, the abortion industry is determined to make money on abortion procedures, even procedures that are not in the best interest of the health of the mother. Researchers with the *British Medical Journal of Sexual and Reproductive Health* likewise assert that, "There is no evidence that receiving medical abortion treatment causes an adverse effect on an ectopic pregnancy" (Duncan). Alternatively, research published in 2021 found that, in fact, "A woman is 30% more likely to die from a missed ectopic while undergoing chemical abortion than if she had not chosen an abortion" (Reardon et al.). The World Health Organization clearly states, "Mifepristone and misoprostol do not terminate ectopic pregnancy." "Even if a pregnancy is ectopic, a woman can experience some bleeding after taking mifepristone and misoprostol because the decidua may respond to the medications" (Clinical Handbook 28). So, not only does mife/miso not complete the abortion, but it can give the woman a false sense of security. Thinking that she has bled out the pregnancy, she could easily continue on with her life, not knowing that the ectopic pregnancy is continuing to grow and become more problematic, potentially even causing fatal hemorrhaging. In summary, an ultrasound is a really helpful tool in gaining knowledge about a pregnancy and making accurate health assessments, but because of its pro-life ties the abortion industry tries to discredit its worth.

In non-ectopic cases of first trimester diagnosis, you can use the same logic as in the second trimester. Since it is impossible for us to have the knowledge of medical experts regarding the maternal risks of every single disease, more general talking points are helpful when

discussing life of the mother cases in the second trimester. As discussed above, abortion is the direct and intentional ending of the
unborn, innocent, human life. Treatment of a maternal fatal disease
is not an abortion, so long as the fetus is not being directly attacked
and the treatment is intended to save the life of the mother, not to
harm the fetus. This would be an appropriate place to reference the
doctors quoted above, and the USCCB distinctions. In the second
trimester, it is important to consider the possibility of saving both
lives, maternal and fetal. It is during this stage of pregnancy that preterm babies are able to potentially survive when given appropriate
care in the NICU. If the mother could stay pregnant for a couple
more weeks, just not the full duration of the pregnancy, then bed
rest, hospitalization, steroid shots, other medications and IVs, can
sometimes be used to buy time and stabilize the baby, all while thoroughly monitoring the situation. A woman facing a diagnosis during
the first week of the second trimester is in a drastically different position than a woman facing a diagnosis during the last week of the
second trimester, when the baby has a very high chance of living
(Abortion Consent 6).

The probability of viability continues to increase in the third trimester. At the same time, the corresponding abortion procedures
become more involved, since the fetus and placenta are much larger
and more developed. This is a good time to break and go review the
abortion procedures explained previously. In the third trimester,
there is a largely sized fetus and placenta that is going to have to
come out one way or another. She can either have labor induced and
deliver vaginally or by c-section. If she has an abortion, she goes
through a nearly identical process. Her labor can be induced and fetal demise occurs before delivery. Or, in rare circumstances, she can

have a hysterotomy abortion, essentially a c-section of a deceased fetus. The big difference with either of these options is whether or not feticide is administered. Whether an abortion takes place or not, a large physical toll is still taken on her body. In some way or another, she will be delivering the fetus/child. So, why would one not induce labor, without feticide, and then care for the preterm baby in the NICU, while simultaneously caring for the mother's severe condition?

Aside from the rare cases of an extreme maternal health diagnosis, the pro-choice side overemphasizes the risk of death of carrying a normal, healthy pregnancy to term. Drumming up fear, they terrify women into believing that they are currently at a greater risk for death if they carry their pregnancy naturally to term, than if they opt for an abortion. On a global scale, "Between 2000 and 2020, the maternal mortality ratio dropped by about 34% worldwide" (Maternal Mortality WHO). This is a huge success that should not be understated. Further, "Almost 95% of all maternal deaths occurred in low and lower middle-income countries in 2020" and "most could have been prevented" (Maternal Mortality WHO). While it is tragically sad that so many women died unnecessarily, it is not accurate to equate their experiences with the experience of a woman in the United States, where exceptional health care is available. Why then is the CDC reporting that pregnancy-related mortality is increasing in the United States through 2021 (Maternal Mortality Rates)? It certainly would be alarming if maternal deaths were substantially decreasing globally while increasing in the United States where advanced medicines and technologies are readily available. This discrepancy could be affected by several factors. Firstly, the United States has a more expansive definition for "pregnancy related death" that includes the time span for one year after birth (Pregnancy Mortality Surveillance 2024). On the other hand, the World Health Organization limits the time span to 42 days after birth, which is the typical postpartum period (Maternal Death). The NIH government website also gives this peculiar definition, "Maternal mortality usually results from a pregnancy, delivery, or postpartum complications; a chain of medical events started by the pregnancy or delivery; the worsening of an unrelated condition because of the pregnancy or delivery; or other factors" (What Examples). Note the words "usually" and "other factors." What does this mean? It is a very unclear definition that compounds confusion. Not only are these definitions different, but the coding for reporting has changed. There have been "changes in the way causes of death are coded, and the addition of a pregnancy checkbox to death records" (Pregnancy Mortality Surveillance 2023). Further, "errors in reported pregnancy status on death records have been described, potentially leading to overestimation of the number of pregnancy-related deaths" (Pregnancy Mortality Surveillance 2023). These factors give confusing, incomplete and even incorrect information, making logical analysis of maternal deaths unavailable.

(Fascinatingly, Donna Hoyert Ph.D. released a report with the CDC about maternal deaths, showing a decrease in 2022 and 2023. While this provides hope, it is worth noting that the maternal death definition used here seems to limit cases to 42 days after the end of the pregnancy, whereas the Pregnancy Mortality Surveillance System of the CDC uses a definition that goes through 365 days after pregnancy) (Hoyert) (Pregnancy Mortality Surveillance 2024).

Another important element to note of the CDC's and WHO's definitions of maternal deaths is that they account for any pregnancy/postpartum related death regardless of pregnancy outcome (Pregnancy Mortality Surveillance 2023) (Maternal Death). While the pro-choice side will have you believe that the maternal death count is for women who carry their children to term and deliver, the actual count is not limited to these cases. Unfortunately, "the denominator of live births used in the maternal mortality ratio reinforces the mistaken notion that all maternal deaths are consequent to a live birth" (Studnicki et al. 2019). In actuality, deaths from abortion complications are also included in the count. However, it is hard to identify how exactly abortion is affecting these numbers since abortion reporting, in general, is not federally regulated and because abortion complications could easily be coded in other categories. For example, if an abortion caused a woman to die by hemorrhage or infection, then hemorrhage or infection could seemingly be listed as the cause of death. By choosing to list the specific symptom that caused the death, they can turn a blind eye to the root cause of that symptom. "Inadequate methods for identifying induced or spontaneous abortion complications assure that most maternal deaths associated with those pregnancy outcomes are unlikely to be attributed" (Studnicki et al. 2019). Even back in 1971, pro-choice advocates "disregard[ed] 7 abortion deaths in New York City on the unsubstantiated theory that these 7 were not performed '... under legal auspices'" (Motion). Essentially, because it was inconvenient for them to acknowledge the high abortion death rate, they simply dismissed and ignored it. More recently, it is also interesting that, "the cause of death is unknown for 7.0% of all 2017-2019 pregnancy-related deaths" (Pregnancy Mortality Surveillance 2023). This seems to be a very large percentage of women dying for no known/documented reason. So, how many maternal deaths are associated with abortion? Research focusing on record linkage found, "an elevated risk of death following abortion, with approximately 20 maternal deaths per 100,000 within 180 days and 34 to 83 deaths within one year, significantly higher than the risk of death associated with childbirth" (Reardon et al.). In contrast, "Without record linkage, only 1% of deaths following abortion will be identified through death certificates alone" (Reardon et al.). Overall, the current systems for reporting and analyzing maternal death trends are filled with severely detrimental shortcomings that lead into fear and short-sighted conclusions.

Time and time again, the pro-choice movement capitalizes on fear. Fear of death, fear of failure, fear of the unknown, fear of anything. They use tired, old scare tactics to lure women into their traps, rushing her through the process and dissuading her from seeking hopeful alternatives. For at the end of the day, the abortion industry is big business. The abortionists and activists line their own filthy rich pockets, while the women who get the abortions are left struggling financially, physically, and emotionally. Women deserve better than to feel coerced into abortions, as if their backs are against the wall and there is no other option. As the pro-life movement, we are called to be a light dispelling the shadows of despair and panic. We are called to walk alongside these women, help them find tangible solutions, and show both them and their unborn children love. Each situation will have different challenges, but each mother deserves love and support. When she feels loved and supported, then she is much more able to love and support the little life growing inside of her. Each of us will have unique opportunities in our lives to live out the pro-life values. No one person can do everything, but God uses even our smallest efforts for the good. When we all do our best, we create a beautifully well-rounded pro-life culture of words, actions, and prayers.

Appendices

Appendices

Metaphors

Violinist argument:

The violinist argument is very popular in college philosophy classes. While there are slight variations the basic argument goes like this: An internationally acclaimed violinist suddenly comes down with a very serious disease and needs a blood transfusion for 9 months or else he will certainly die. However, it cannot be a blood transfusion from just anyone, you are the only person in the whole world whose blood matches. Knowing this, some of his fans sneak into your house, knock you unconscious, bring you to the hospital and hook you up to the blood transfusion machine. When you return to consciousness, the doctor tells you that you have to stay hospitalized and hooked up to the violinist for 9 months or else you will kill him. What is stronger: the violinists right to life or your right to bodily autonomy?

While, at first glance, this metaphor may seem intimidating, longer analysis finds many flaws in the comparison.

1.	Firstly, in the vast majority of cases, the mother has chosen to participate in sex, knowing that sex can and often does lead to pregnancy. Yes, there are some very tragic instances of rape, but those only account for 1% of abortions (Finer et al. 113). This causes the pro-choice to include women whose contraceptives failed, in this category. They claim that since the woman was actively trying to prevent pregnancy, that the pregnancy occurred completely against her consent. However, contraceptive failure is entirely different from rape. The

175

inconsistencies of contraception effectiveness has been well documented and is available for the public to read. Thus, the woman knows that, while she may be decreasing her chances of getting pregnant, her actions are still consenting to doing the very thing that always has the possibility of ending in pregnancy. Therefore, consensual sex, even when contraceptives are used, is incomparable to being knocked unconscious and taken to a hospital.

2. Secondly, there is the issue of parental obligation. Parents have a greater responsibility towards their own children. For example, choosing to feed starving children in Africa is a very noble thing to do, but is not a requirement and you will not receive punishment for failing to donate food. However, you are responsible if your own child is starving and you will certainly face the consequences. Parents have a natural obligation to protect and provide for their children, regardless of whether or not they feel like it. This is why fathers have a responsibility to pay child support, even if they did not want to be a father and even if they were using contraception when having sex.

3. Third is the natural use of organs. The inherent purpose of the uterus is to nurture new life into the world. One's blood is inherently designed for the use of its owner. While it can be used to help other people, that is not its original function. The primary purpose of the uterus, and the rest of the woman's reproductive system, is to reproduce. Women certainly do not shed their uterine linings every month because it is fun. Rather, they have a period every month because

their body is preparing to be a hospitable environment for their child.

4.   The fourth difference is that of ordinary measures. Pregnancy is a normal and healthy stage of human development. While it is unpleasant for a woman to have morning sickness, there is nothing inherently wrong with her body. In the case of the violinist, being hooked up to a blood transfusion machine for 9 months is beyond ordinary care. One could have a separate debate about whether staying plugged in would be the virtuous thing for the person to do, but it is not indicative of the reality of pregnancy. Pregnancy is very natural. It only requires that we give the fetus standard nutrition and its normal environment. Regular pregnancy does not require extraordinary care.

5.   Similarly, being pregnant for 9 months is extremely different from being hospitalized for 9 months. Pregnant women hold jobs, work out, visit with family, go where they want to go, eat normal food and do plenty of other everyday life activities. The hospitalized person in the example is severely restricted. Yes, some women do experience pregnancy complications which limit what they can do, and even sometimes have to go on bed rest. However, these bedrest cases are the exception. That is not what normally happens in a healthy pregnancy and does not take over the full 9 months.

6.   Lastly, and most importantly, the cause of death and the corresponding intention are critically important. Abortion is the direct and intentional killing of the unborn life. Removing a plug for a blood transfusion is removing extraordinary

care. While it would, theoretically, most likely lead to the violinist's death, you are not doing anything to directly kill him. A more apt comparison would be the hospitalized person administering poison to the already vulnerable violinist. Abortion is the direct killing of the child.

After going through these responses, try to reorient the conversation back to abortion. You do not want to spend the entire debate talking in code. People typically use these metaphors to ignore the reality of abortion and instead philosophize on theories.

Burning Building Metaphor:

Pro-choice Argument: You are the only adult in a fertility clinic when a fire breaks out and is rapidly consuming the entire building. (Already completely implausible.) As you run out the door, you can grab a crying 5yo who was left in the corner or a container of 1,000 embryos. (Don't ask why the 5yo was left there or why you couldn't grab one with each hand.) Who do you choose? If you don't choose the 5yo, you are a terrible, cruel human being; and if you don't choose the 1,000 embryos, then you don't really believe that life begins at conception.

First off, this situation is a lose - lose situation, but it is also completely absurd and is never really going to ever happen. Further, some people might freely wish to do a third option and rescue the child and then risk the fire again to rescue the embryos. Also, this scenario completely discredits automatic fire sprinkler systems, quick response times by first responders, location of how close to the exit you, the child, and embryos are, and the severity of the fire. Keep in mind, we do not have to acquiesce to all of their bizarre stipulations on a made up situation. All that being said, if someone chooses to save the 5yo life, instead of grabbing the container of embryos that would be a very understandable and good response. The 5yo clearly has a very high chance of continuing to live, whereas the embryos are burdened by many factors that make it more difficult for them to reach birth and life outside the womb. One of which is temperature. If the building is burning or if the embryos are frantically being moved, then the temperature of the container and levels of nitrogen, freezing the embryos, would almost certainly get messed up. This would very likely end in the embryos' demise. Building on this train of thought, the toddler only needs ordinary care - the removal from the fire - to stay alive. The embryos are going to be dependent on the removal from the fire and on many forms of extraordinary care and medical interventions in order to remain alive. The emotional response is also to rescue the 5yo, who is already very bonded with his/her family, can acutely experience pain, is terrified, and is crying out for help. It would be wrong to ignore his/her plea for life saving aid. The toddler's life is very clearly valuable. Though many try to ignore it, each embryo's life is also very valuable. All of the lives have value. Saving the life of the 5yo does not somehow take away the inherent value of the embryos' lives. For example, if the situation were slightly different and you could save either your own child or 10 other children, most everyone would choose to save their own child. We, as parents, have a naturally greater responsibility to take care of our own children. The parent who rescues his/her own child is not denying the humanity of the 10 other children, (s)he just has limited resources for rescuing. That being said, there is also an immeasurable difference from saving one life and being unable to save others, from actively destroying those others. If one wishes to save people, but is unable to do so, that could not be more different from one who is actively trying to cause their death. Abortion is the direct and intentional ending of innocent preborn life. Evacuating a burning fertility clinic is not.

Alien Metaphor:

Pro-choice Argument: If tiny fetuses have a right to life, do aliens also have a right to life (if they were to come and invade earth)?

Essentially, by bringing this up, they are trying to make you sound ridiculous for advocating for alien rights and are also trying to obscure the heinousness of what abortion actually is. While you do not want to get dragged down a rabbit hole of nonsense, you also do not want them to think that they came up with a stumper question that you do not have a response to. So, it is best to give a short, concise answer and then reorient the conversation back to discussing abortion.

- Aliens are not human. There would be physical, mental, emotional, and psychological differences.
- They do not have an immortal soul, will and intellect. They would not be made in the image and likeness of God.
- Therefore, they would fall more in the animal category. While protecting alien life might be the kind hearted thing to do, it is on an entirely different moral sphere as aborting human babies in the prenatal stage.

How to Help a Pregnant Woman

While all of the above talking points are true and extremely important, they do little good when one is faced with an actual crisis pregnancy. All of the sudden, viewpoints and beliefs get ousted by the reality of the situation. And the reality of the situation for nearly all women in a crisis pregnancy is that they are panicking about being pregnant when they really, really do not want to be. Logic and past values easily get thrown out the window and replaced by very strong emotions, feelings and physical sickness. Because of this change in mentality, along with the increased gravity of the situation, a different approach is warranted. The immediate goal is to calm, comfort, and care for the mother and help her to choose life for her child. This is different from the goal of debates or other conversations where you are trying to change the other person's entire perspective on the pro-life movement. In this pivotal moment, that one, specific, unborn life is of the utmost importance and everything you say to the mother needs to encourage her along the path of choosing life for that child, regardless of whether or not she is ready to commit to being ideologically aligned with the entire pro-life movement.

Life throws us all kinds of curve balls, but one of the most unexpected and shocking conversations can be when a friend, or acquaintance, confides in you about an unplanned and undesired pregnancy. The flood of emotions and shock can hit really hard. You might have been completely unaware that she was having sex. Maybe she had previously identified as pro-life and you never thought she would be considering abortion. Maybe you just became friends recently and are still getting to know her. Maybe you do not know if she has told anyone else, what the father's opinion is, if her religion will have an effect, or what her biggest obstacles are. Essentially, you might be confused and have every question in the book, and not know where in the world to start the conversation! Don't fret, the following five steps are guidelines to help you navigate this tricky conversation and the following events.

Step 1: Listen. If you are a little shell shocked about the news and need a minute to process, that is okay! One, simple, kind remark and the invitation for her to share her story is all you need in the moment. She clearly has a lot on her mind and came to you to talk about it. So, she should welcome the opportunity for her to talk it out and for you to just be with her in that moment listening to her. This gives her the much needed chance to share her heart and you the much needed minute to collect your thoughts and get a feel for the whole of the situation. If you are unsure what to say in your initial response, here are a few options: "I'm so glad you told me." "Do you want to go somewhere more private to talk about it?" "Ok, I'm here for you." "I love you and I'll help you get through this." "You're probably going through a whirlwind of emotions, walk me through it." Give her a hug. Offer her a tissue. Do what feels natural and right based on your relationship with her. You might also say, "Let's just take a minute and chat, you don't need to rush into any decision." Time is on your side. She needs time to calm the intense emotions and view the situation with more clarity and strength. Once she has agreed to talk more, listen to her whole story. Make a mental note to yourself that you are in this for the long haul. Give her as much time as she needs. Helping her in this pivotal moment is truly the most important thing you can do that day.

Step 2: Pinpoint her biggest obstacles. These obstacles are the driving force behind her intent to abort. Before we can change her intent for abortion, we have to help remove her obstacles and before we remove those obstacles we have to know what they are. How do we identify the obstacles? Listening. Listening and paying attention to three elements: repetition, emotionalness and/or a list. If she lists off her obstacles, then that makes it extremely easy to identify what is going on. However, oftentimes she will be so overwhelmed that she does not have the clarity of thought to recognize what her challenges even are. This necessitates a keener sense of perception on your part, looking for repetition and heightened emotions. If there is a certain element of her story that she keeps coming back to, that signifies its importance. For example, if she keeps making comments like, "the dad is such a loser," "he would be such a terrible dad," or "I never want to speak to him again;" then clearly the father is a big negative factor. Or, perhaps, she only makes one comment about the dad but starts tearing up and has a shaky voice while doing so. This also reveals that he has a large impact on her decision. Even if these matters seem comparatively small to you, if she keeps bringing them up or becomes really upset when talking about it, then they feel very large to her and need to be thoroughly addressed. After you have identified one obstacle, even if it is a large one, be on the lookout for more. Guttmacher Institute found that women gave a medium number of four reasons and as many as eight, when asked why they were pursuing an abortion (Finer et al. 113). Life is complicated and these women are much more likely to choose life when we acknowledge her challenges and find real tangible ways to help her forward.

Step 3: Take concrete steps forward. This is another way in which these pregnancy conversations differ from typical abortion debates. Debates exchange words, ideas, and philosophies. Women in crisis pregnancies need reassuring words, ideas for a path forward, and help following that path. We need to figuratively, and sometimes literally, take her hand and lead her along the way, providing hope and support. Oftentimes, she can easily be way too overwhelmed to take action on her own. The extreme stress that she is in can lead to paralyzing fear. Even calling a doctor might terrify her to the point of not being able to dial the number. This is where you come in on a very practical level. Help her make the contacts she needs; community resources, a pregnancy center, and/or a doctor. Depending on her specific needs, maybe she wants you to make the phone call, go with her to initial appointments, or maybe just be on the phone with her afterwards so she can process how it went. Every woman is unique and so the way that you care for her and help her in the moment is going to be unique.

Helping her get an appointment at the local (pro-life) pregnancy center is a really good place to start. Those centers often have free ultrasounds, are very good at encouraging pregnant women, and have a wide knowledge of local resources that can help alleviate her other obstacles. Heartbeat International has a worldwide directory of pregnancy care centers. All you have to do is plug in your location and it will tell you the closest centers, their contact info and the services that they offer.

Another really good initial contact is a pro-life doctor. The doctor can speak from a position of authority and medical expertise and, thereby, help alleviate any medical fears that she has. If you are unsure which doctors in your area are pro-life, then look for one listed on the American Association of Pro-life OBYNs directory (Find OBGYN). You can also search for one that is NaPro accredited or who specializes in Restorative Reproductive Medicine (Naprotechnology) (About IIRRM). While doctors do not typically list "pro-life" in their bios, they do list their accreditations. So, if you find one that is NaPro accredited, then you know that (s)he will for sure be pro-life. Lastly, you can try searching for a doctor who is part of the Catholic Medical Association (Catholic Physician). A doctor who is part of this association is more likely to be pro-life. Even if the patient is not Catholic, she can absolutely still receive care from one of these doctors. However, be conscientious of how you approach the topic and avoid saying something like, "Oh my gosh I have the best Catholic doctor ever who is NaPro and totally pro-life and will totally convince you to choose life!" That is going to do little good in helping her move forward. Instead you can say something like, "I know this well-respected doctor in the area who specializes in pregnancies occurring in tricky situations, would you be interested in talking to her/him?"

Whether an ultrasound takes place at a pregnancy center or a doctor's office, it is one of the most effective ways to help a woman choose life. Seeing the baby move on the screen and hearing the heart beating on the doppler, humanizes the baby. It changes the frame of reference from an abstract idea or just a blob of tissue, to a moving, living baby. Because of this, it is of the utmost importance for her to have an ultrasound. In some very tricky situations, the woman might be dead set on abortion and unwilling to have an ultrasound because she views it as a pointless waste of time. In such a scenario, a version of the following conversation might be appropriate:

Her: "I'm not going to get an ultrasound. It'd be a waste of time and I already have my appointment at Planned Parenthood next week to get the pills."

You: "Do you know how far along you are?"

Her: "Not really, I took the test a few weeks ago."

You: "Okay, well you might not even qualify for the pill type of abortion if you're too far along. What if you got an ultrasound to check how far along you are? More knowledge is always helpful."

Or

Her: "Ugh, I can't wait for next week. I'm already spotting and I just want to get this whole thing over with!"

You: "You're spotting? Miscarriage is pretty common (10% to 20%) in the first trimester (KFF miscarriage). What if you got an ultrasound to see if you are miscarrying? If you're miscarrying naturally, then an abortion isn't an option and going to the clinic would be a waste of money."

Or

Her: "I'm having some really weird symptoms. My appointment can't come fast enough!"

You: "Weird? What do you mean? It might be helpful to see a doctor and get an ultrasound done, so that you know exactly what's going on and what's causing those symptoms. An ultrasound can check to make sure that the pregnancy is not ectopic.

Because if it is, then you need specialized treatment and an abortion could actually make matters worse for your health."

Aside from medical support, it is also very important to help her find resources and support for her other areas of stress. Whatever her obstacles are, find ways to help her check them off the list. This might mean helping her sign up for Medicaid, looking into pregnant student scholarships, finding affordable and child friendly housing, finding an adoption agency, throwing her a baby shower or many other things. There are so many different ways in which you can help! Know your strengths, work with community support, and simply do the best you can to authentically love and care for her!

Step 4: Be with her every step of the way. If you are doing the above steps, chances are pretty good that you have already been walking with her every step of the way. However, if perhaps she is more independent and does not need much help past the initial conversation, then do not forget to still check in on her. Whether you text, call, or meet up with her in person, it is important to continue to show authentic care for her. We are not there only to save the baby. While it is of the utmost importance that the baby lives, it is also of the utmost importance that the mother feels supported and empowered in her new phase of life. Not only is checking in on her the right thing to do, it is also the most effective thing to do to continue to make sure that she does not abort. Volatile emotions and ever changing situations are very real and can make a pregnant woman change her intentions about the pregnancy in an instant. Staying in touch with her and continually providing more support, helps to make sure that she does not suddenly change her mind. In this way, it is both a friendly and effective tactic.

Step 5: Later goals. It can be very tempting to have a whole bunch of goals and to try to help her achieve them all at once. However, this is oftentimes not the most effective method. If she feels overwhelmed, she will likely retreat and shut you out; thereby disabling you from having any positive influence. Because of the gravity of the situation, it is important to make sure that she chooses life for that one baby. Keeping that one baby alive, and helping the mother through this, is of paramount importance. Once the worry of abortion has passed, then other goals can be focused on. The specific timeline and order will vary with each woman. Next goals could be abstinence until marriage, becoming fully ideologically pro-life, and being reconciled with her faith. Regarding abstinence, hopefully she is so focused on the pregnancy that, if she is single, she is not continuing to have sex. You could gently point out that abstinence will 100% guarantee that she does not have this type of crisis situation again, and that marriage has a unique beauty and ability to bond the spouses together in loving fidelity. Regarding being wholly pro-life, it is unlikely that she is going to instantly switch from actively considering abortion to agreeing wholeheartedly with the pro-life movement. Knowing that, give her time and have lots of conversations with her, slowly alleviating her fears and misconceptions and replacing them with knowledge of what the pro-life movement really does stand for. Also, it is extremely beneficial for her to start practicing her faith again. If she is Protestant or Jewish, talking with her pastor or rabbi is a great place to start. If she's Catholic, talking with the parish priest is likewise a good starting point, which can then lead to the Sacrament of Confession and full reconciliation with the Faith. However, the timing really is up to her. Confession cannot be forced. It needs to be a free act, with true contrition (which is not probable if she is still actively considering abortion - which has the matter of being a mortal sin). So, ultimately, it is her decision. All you can do is help lead her towards the truth. Again with all of these goals, there is no firm timeline. Use your discretion about when and how to best approach these new goals.

If you are talking to the father, instead of the mother, you can still follow the above outline. You will just want to tweak it slightly to be applicable to the parent who is not actively pregnant. Overall, the same practices of patient listening, addressing problems, and providing support and encouragement, will stay the same.

Adoption

Adoption frequently gets a bad rap because it is widely misunderstood. Additionally, it is often confused with the foster care system. Even the supreme court justices lied about adoption in their dissent from *Dobbs v. Jackson*. They claimed that adoption does not diminish the "financial costs of going through pregnancy and childbirth" (*Dobbs* Dissent 40). This, along with many other myths surrounding adoption, is simply not true. While there are flaws to be found in both the foster care and private adoption systems, there are several key differences between the two. Explaining the differences and the actual process of private adoption can help alleviate the fears surrounding that option.

Freedom to Choose:

- Foster care: Children are removed from their parents and their home without any say so from the parents. DCS can come to the home and forcibly remove the children without parental consent.
- Adoption: The birth mom (and dad) have the freedom to choose adoption. Nobody is removing the child from her care without her prior knowledge and complete consent. She has freedom to change her mind about adoption and change her plans while she is pregnant and up until she signs the official paperwork. (The timing of the signing of the paperwork changes with each state, but is often at least 24 hours after birth or later.)

The couple:

- Foster care: When DCS places a child in foster care, the foster care agency chooses the couple with whom the children will be placed.
- Adoption: The birth mom (and dad) choose the exact couple with whom they want to place their child. She can be as picky as she wants and can disqualify a couple from her consideration for any reason. She can choose a couple based on a certain education level, marital status, matching values, race, religion or anything else that is important to her.

Contact:

- Foster care: When children are taken into foster care, the mom does not get to decide what level of contact (if any) that she has with the children. The children were removed from her care because of a grave concern, so the foster care agency / DCS decide what level of contact they believe would be safe and beneficial.
- Adoption: The birth mom (and dad) choose the level of contact that she wants during the pregnancy and then in the following years. She can choose to meet and interview the couple before choosing them. She also decides if she wants a closed, semi-open or open adoption and then thereby picks a couple who is agreeable to that contact level and will honor her wishes after they gain custody of the child. (The legality of official post adoption contracts vary by state.)

Cost:

- Foster care: The birth parents can incur legal fees when fighting to regain custody in the courts.
- Adoption: While the specifics vary by state, the adoptive couple covers most, if not all, of the related expenses. Typically, the adoptive couple will pay for the medical costs of the pregnancy and the hospital bill (though Medicaid frequently helps with that), as well as fees for the attorney(s) and adoption agency. Often, the adoptive couple will also pay for some counseling for the birth mom, as well as some living expenses.

Counseling and Support Groups:

- Foster care: Birth parents do not receive counseling before a child is removed from their care and placed in foster care.
- Adoption: The birth mom often receives free counseling before the birth of the child in order to prepare her emotionally and support her along the adoption journey. There are many free agencies and support groups who specialize in supporting the women for years after the birth and adoption placement.

Judgment and Public Perception:

- Foster care: The removal of children from one's home is rather public and a hard fact to hide. People will quickly notice if someone no longer has their child with them and an explanation can be embarrassing and painful. Children are placed into foster care when the circumstances are dire, so people will automatically know that the state does not consider that person to be a fit parent, the living situation unsafe, or something along those lines.

- Adoption: The choice to place a child in an adoptive home is made freely and is not forced by the state because of abuse or neglect. Placing your child for adoption, in no way automatically labels a birth mom as an unfit parent or one who is unable to properly provide for a child. Because of this, they are less susceptible to people making harsh judgments about them. Choosing adoption is not an act of defeat, but rather an act of great courage and self-sacrificial love.

Child's Perspective:

- Foster care: Children in foster care have usually experienced abuse or neglect. Frequently, they are also older and can have strong memories of their original family. They may have contact with their birth parents completely cut off. These, along with other factors, can add up to a wide range of emotions and feelings about their birth parents.

- Adoption: Commonly, children are placed as infants with their new adoptive family and so they grow up forming memories with that one family from the very beginning. Also, they have not endured abuse or neglect and so do not have residual trauma from that. The adoptive parents have a really important role in helping the adopted child grow up loving the birth mom and appreciating the many sacrifices that she has made. With open and semi-open adoptions, the children can have some level of contact with the birth mom and can experience her love and devotion first hand.

Custody:

- Foster care: Children in foster care are under the custody of the state. Therefore, the foster parents have somewhat limited control over how they parent the children.
- Adoption: At finalization, the adoptive parents are the legal parents of the child and have full custody. This enables them to make all of the parenting decisions. Even before finalization, the adoptive couple still has much greater control, in regard to parenting decisions, than foster parents.
-

All that being said, there are so many loving foster families who provide a beautiful, safe and caring home for children in foster care. Whether these children are with them temporarily or become permanently adopted, the foster parents provide them with an outpouring of love. Their ability to do so, despite all the logistics and control of the state, is a true testament to their character. Similarly, couples who go through the process of private (often infant) adoption also love their children with their whole heart. These children are an answer to prayer and build beautiful families. There is a mutual gift of self. The children are a blessing to the parents, and the parents are a blessing to the children in turn.

Overall:

- Adoption should always be chosen freely by the birth mom without any undue outside pressure.
- She should be made fully aware of all of her rights through-out the entire process.
- While adoption is certainly a thousand times better than abortion, adoption should always be talked about gently and with the utmost care. It should not be pushed as a "quick fix," because it is a very involved and emotional process for all included. Since many people are quick to dismiss adoption, a little understanding of the challenges that accompany adoption can go a long way in making the other person feel heard before going into an explanation of the benefits of adoption.
- If a woman desires to parent the child, all reasonable support should be given to her to help her raise the child.
- Adoption is best talked about in situations where the woman very much does not want to raise a child. In situations where she is open to raising the child but is worried about medical complications with the actual pregnancy, finances or such; addressing the root cause of her concerns is the better re-sponse. Otherwise, she can quickly feel like you do not care about her and her feelings and only want the baby.
- The wording about adoption in a debate will vary slightly from the wording one would use when talking directly to a pregnant woman considering adoption, but the basic princi-ples remain the same.

Adoption Agencies:

Whether you are a birth mom, adoptive couple, or a helpful friend, when looking for an adoption agency look for one that prides itself on its integrity and ethical practices. Just like any business, some are run much better than others and you will need to associate yourself with the best. An exhaustive list of agencies is too much to list here, so a good way to start is by looking at small to medium sized agencies in your state, particularly ones with a faith background. When the child has a medical condition that qualifies them for a special needs adoption you can specifically look at these U.S. agencies that work regularly with couples prepared for the extra logistical challenges.

MLJ Nightlight Christian Adoptions
Lifeline Children's Services
The Cradle
Spence Chapin Adoptions
National Down Syndrome Adoption Network

Birth control

While the reversal of *Roe v. Wade* did not alter access to birth control in any way, it is essential to understand how closely intertwined birth control and abortion really are. To begin, birth control gives women a false sense of security that they will definitely not become pregnant. However, birth control methods continuously fail, leaving these women with unintended pregnancies. Since these women have already adapted the mindset that it is okay to forcibly alter their body's natural hormones and reproductive cycle, or block man's sperm which is the natural result of intercourse, it is pretty easy to understand how abortion seems to be the next logical step for them. They have been actively avoiding having a child for so long, that an abortion can be seen as another way to avoid having a child. They turn a blind eye to the humanity of the person who has already been conceived and treat abortion as a sort of backup to their regular form of birth control. This is so common that, "Over 50% of patients having abortions used a contraceptive method during the month they became pregnant" (Hufbauer et al. 9). In this way, abortion has become their Plan B or Plan C.

With Plan B, typically being the abortifacient Plan B pill, abortion is so commonly becoming known as Plan C that there is now a whole website entitled *Plan C*. This website advises women how to get abortion pills in the mail, how to circumvent your state's restrictions and if pills from India are available. It even goes so far as to advocate for ordering pills in advance, in case you might want an abortion later (Plan C) (Plan C Instagram). This highlights the circular nature of using birth control, its failing, having an abortion, and getting back on birth control.

Let's focus back on Plan A, regular birth control. Birth control can come in several different forms: barrier, IUD, hormonal patch and pills. Barrier methods, such as condoms, work by preventing the sperm from reaching and fertilizing the egg, and are some of the least effective forms of birth control. Male condoms have a typical use effectiveness rate of 82% and female condoms of 79% (Healthwise Staff Effectiveness). This leads to many unplanned pregnancies.

Hormonal methods, such as pills, patch or vaginal ring, have a slightly higher effectiveness rate around 91%, but come with a plethora of side effects and still account for many unplanned pregnancies (Healthwise Staff Effectiveness). For example, Mayo Clinic lists the side effects of the patch as, "Breakthrough bleeding or spotting / Skin irritation / Breast Tenderness or pain / Menstrual pain / Headaches / Nausea or vomiting / Abdominal pain / Mood swings / Weight gain / Dizziness / Acne / Diarrhea / Muscle spasms / Vaginal infections and discharge / Fatigue / Fluid retention" And a greater risk for: "Blood clotting problems, heart attack, stroke, liver cancer, gallbladder disease, high blood pressure" (Birth Control Patch). Years of hormonal manipulation and even a couple of these complications are severely detrimental to the woman's short and long term health and well being. Further, the way in which they prevent pregnancy is much more complex. A hormonal method can work a number of ways such as it, "Prevents ovulation / Thickens mucus at the cervix so sperm cannot pass through / Changes the environment of the uterus and fallopian tubes to prevent fertilization and to prevent implantation if fertilization occurs" (Healthwise Staff Prevent). This last method, preventing implantation, makes these hormonal methods abortifacients.

A type of birth control is an abortifacient if it prevents a tiny unborn child, who has already been conceived, from implanting in the uterus, as is normal. If the blastocyst (unborn child), is unable to implant in the uterus, (s)he will die and will be excreted from the mother as if she is getting her period. Since implantation did not occur and she could not take a positive pregnancy test, the woman will not be able to know if she had conceived a child. So, she will not be sure which way the birth control worked and whether or not it utilized its abortifacient capabilities. The USCCB addresses the morality of such abortifacients and clarifies that, "Every procedure whose sole immediate effect is the termination of pregnancy before viability is an abortion, which, in its moral context, includes the interval between conception and implantation of the embryo." (Bransfield 18) However, birth control companies attempt to hide this function of their contraceptives and deny the abortifacient potential by changing definitions, yet again. Planned Parenthood gives the definition, "Pregnancy officially starts when a fertilized egg implants in the lining of the uterus. It takes up to 2-3 weeks after sex for pregnancy to happen" (How Pregnancy). The American College of Obstetrics and Gynecology essentially agrees and states, "intrauterine pregnancy begins when a fertilized egg implants itself in the uterus" (ACOG Guide). By willfully ignoring the pre-intrauterine stage of pregnancy, when the zygote/blastocyst is in the fallopian tube, these companies are woefully misguiding women.

While Plan B, the morning after pill, is the most widely known abortifacient, the hormonal methods above and the IUD also can act as abortifacients, ending the life of an unborn child before anyone even knows that (s)he exists. IUDs, which can be either hormonal or copper based, either prevent fertilization or implantation (Healthwise Staff Prevent). Planned Parenthood goes so far as to claim that their Paragard IUD is more than 99.9% effective, so long as you use it within 5 days of having sex (IUD). However, they do not clarify what part of her cycle in which the woman would need to be. Therefore, the woman could easily have had sex while ovulating, conceived, and then inserted the IUD days afterwards. Clearly, in this case pregnancy would not have been prevented, rather a very early term abortion would have taken place. (If a pregnancy does make it to the implantation stage despite the IUD, there is "a high likelihood of implantation in a tube (ectopic) and the foreign body dramatically increases the risk of infection") (Reardon et al.).

On a similar note, "menstrual regulation" is being brought into the discussion now. Essentially, a woman is late for her period and is prescribed misoprostol, and sometimes mifepristone, without ever having taken a pregnancy test. She then gets her "period" without ever knowing for sure if she had been pregnant. The FDA also discusses "menstrual regulation" occurring by a technique identical to aspiration abortion (CFR). Similar tactics date back to the 1970s before abortion was widely available (Edström 490). Now that *Roe* has been reversed, "menstrual regulation" is coming into the forefront again in an attempt to skirt around the law and still have access to abortion (Lenharo).

Whether a new life comes to an end via an abortifacient or menstrual regulation or by a legalized abortion, the tragic end result remains the same. A precious, innocent life was forcibly ended much too soon. Birth control and legalized abortions are intricately linked both in function and accompanying mindset. The National Center for Biotechnology Information even lists "Abortion" as a type of birth control (Hatcher and Kowal). Abortion doula Ash Williams, who presents as a trans man, agrees with NCBI and goes a step further advocating for multiple abortions, saying, "Abortion is, for some, a form of contraception. That shouldn't be limited" (Jones). Even the dissenting Justices in the *Dobbs* case noted that,"'[R]easonable people... could believe that 'some forms of contraception' simi-larly implicate a concern with 'potential life'" (*Dobbs* Dissent 24). The cyclical codependency is clear. Plan A is birth control. Plan B is an abortifacient morning after pill. And, when those fail, Plan C is an abortion; after which the woman goes back to Plan A, contraception, and begins the cycle all over again.

Well are there any ways to avoid pregnancy without taking an abortifacient styled contraceptive? What about condoms? While condoms act as a barrier method and do not as an abortifacient, there are moral qualms with their use, since they directly and intentionally interrupt marital unity. Sex is of the utmost intimate nature and the complete gift of oneself to one's spouse. In a way, using condoms is lying with one's body. Through sex you are telling yourself that you are giving your full self to your spouse, but in reality you are forcibly holding back a very essential element. The man is not fully giving himself to his bride, if he is withholding his sperm from her and the woman is not fully giving herself to her groom, if she is rejecting his gift of self. Imagine that you were invited to someone's home for dinner. Upon entering, they extend their hands for a handshake. You shake their hand, but you make sure to put on a glove before doing so. You do this to "protect" yourself. Your hosts would be surprised and put off by the disrespectful and distancing nature of your actions. So too is using a condom an act of detachment from one's spouse at the very moment that they are supposed to be the closest.

So what can women and their husbands choose to space out births in order to best provide for her well being and the well being of their whole family? Natural Family Planning is an umbrella term that includes multiple individualized methods of determining when a woman is fertile and then abstaining from sex for a short period of time during her window for ovulation. These methods work by noticing trends in her bio markers, Creighton Method, additionally noticing changes in body temperature, Symptothermal Method, and additionally by taking ovulation tests, Marquette Method. None of these methods work as abortifacients or as barriers to marital unity. They simply study the woman's natural cycle and then help couples make educated decisions about when to have sex based on whether they are trying to avoid or achieve pregnancy.

Effectiveness rates (when used as directed):

Marquette Method 98%-99%
Symptothermal Method >99%
Creighton Method >99% (Smoley and Robinson)

Works Cited

"2009 California Health and Safety Code - Section 123460-123468 :: :: Article 2.5. :: Reproductive Privacy Act." Justia Law, Justia, 2009, law.justia.com/codes/california/2009/hsc/123460-123468.html#:~:text=SECTION%20123460%2D123468,-123460.&text=The%20Legislature%20finds%20and%20de-clares,choose%20or%20refuse%20birth%20control.

"2012 Code of Alabama :: Title 26 - Infants and Incompetents. :: Chapter 22 - Abortion of Viable Unborn Child. :: Section 26-22-2 - Definitions." Justia Law, Justia, 2012, law.justia.com/codes/alabama/2012/title-26/chapter-22/section-26-22-2/.

"Abdominal Hysterectomy." *Mayo Clinic*, Mayo Foundation for Medical Education and Research, 28 Feb. 2023, www.mayo-clinic.org/tests-procedures/abdominal-hysterec-tomy/about/pac-20384559.

"Abortion Bans in Cases of Sex or Race Selection or Genetic Anomaly." *Guttmacher Institute*, Guttmacher Institute, 31 Aug. 2023, www.guttmacher.org/state-policy/explore/abortion-bans-cases-sex-or-race-selection-or-genetic-anomaly.

AbortionClinics.Org,Inc. *Kindness, Courtesy, Justice, Love, Respect*, AbortionClinics.Org, Inc., Bethesda, Maryland.

"Abortion Informed Consent Brochure." *Indiana State Department of Health*, Indiana State Government, 29 June 2022, www.in.gov/health/files/abortion-informed-consent-brochure.pdf.

"Abortion in the United States Dashboard." *KFF*, KFF, www.kff.org/
    womens-health-policy/dashboard/abortion-in-the-u-s-dash-
    board/#key. Accessed 5 Feb. 2025.

"Abortion: Last Resort for Hellish Morning Sickness." *ABC News*,
    ABC News Network, 29 June 2010, abcnews.go.com/Health/
    Wellness/women-hellish-morning-sickness-elect-abortion/
    story?id=11043146.

"Abortion Policy." *ACOG*, American College of Obstetricians and
    Gynecologists, May 2022, www.acog.org/clinical-information/
    policy-and-position-statements/statements-of-policy/2022/
    abortion-policy.

"Abortion Reporting Requirements." *Guttmacher Institute*,
    Guttmacher Institute, 1 Sept. 2023, www.guttmacher.org/state-
    policy/explore/abortion-reporting-requirements.

"Abortion Risks: LA Dept. of Health." *Abortion Risks | La Dept. of
    Health*, Louisiana Department of Health, ldh.la.gov/page/abor-
    tion-risks. Accessed 21 Feb. 2024.

"Abortion Surveillance - United States, 2020." *Centers for Disease
    Control and Prevention*, Centers for Disease Control and Pre-
    vention, 23 Nov. 2022, www.cdc.gov/mmwr/volumes/72/ss/
    ss7209a1.htm.

"Abortion Surveillance - United States, 2021." *Centers for Disease
    Control and Prevention*, Centers for Disease Control and Pre-
    vention, 24 Nov. 2023, www.cdc.gov/mmwr/volumes/71/ss/
    ss7110a1.htm.

"Abortion Surveillance - United States, 2022." *Centers for Disease Control and Prevention*, Centers for Disease Control and Prevention, 28 Nov. 2024, www.cdc.gov/mmwr/volumes/73/ss/ss7307a1.htm

"About." *International Institute for Restorative Reproductive Medicine*, International Institute for Restorative Reproductive Medicine, https://iirrm.org/about/. Accessed Jan. 2025.

Abrevaya, Jason. "Are There Missing Girls in the United States? Evidence from Birth Data." *SSRN*, Elsevier Inc., Feb. 2008, papers.ssrn.com/sol3/papers.cfm?abstract_id=824266.

"ACOG Guide to Language and Abortion." *ACOG*, ACOG, Accessed Jan. 2023.

"ACOG Guide to Language and Abortion." *ACOG*, ACOG, www.acog.org/contact/media-center/abortion-language-guide. Accessed Jan. 2025.

Affidavit: Sandra Cano, formerly known as Mary Doe, v Arthur Bolton et al. 2003. https://thejusticefoundation.org/wp-content/uploads/2020/05/Sandra-Cano-Affidavit.pdf.

Ajmal, Maleeha, et al. "Abortion." *StatPearls [Internet].*, U.S. National Library of Medicine, 15 July 2022, www.ncbi.nlm.nih.gov/books/NBK518961/.

Anderson, Jane et al. "Reproductive Choices of Young Women Affecting Future Breast Cancer Risk." *American College of Pediatricians*, American College of Pediatricians, Oct. 2019, https://acpeds.org/position-statements/reproductive-choices-of-young-women-affecting-future-breast-cancer-risk.

Arey, Whitney, et al. "A Preview of the Dangerous Future of Abortion Bans — Texas Senate Bill 8." *The New England Journal of Medicine*, 4 Aug. 2022.

Attia. "How Much Does It Cost to Get an Abortion?" *Planned Parenthood*, Planned Parenthood Federation of America Inc., Nov. 2022, www.plannedparenthood.org/blog/how-much-does-it-cost-to-get-an-abortion.

Auger, Nathalie et al. "Second-trimester abortion and risk of live birth." *American Journal of Obstetrics and Gynecology*, vol. 230, no. 6, U.S. National Library of Medicine, 7 Nov. 2023, https://pubmed.ncbi.nlm.nih.gov/37939985/.

Baglini, Angelina. "Gestational Limits on Abortion in the United States Compared to International Norms." *Charlotte Lozier Institute's American Reports Series*, no. 6, Feb. 2014, https://lozier-institute.org/internationalabortionnorms/.

Baird, D T. "Mode of action of medical methods of abortion." *Journal of the American Medical Women's Association* vol. 55, no. 3, U.S. National Library of Medicine, 1972.

Bartee, Lisa, and Christine Anderson. "Properties of Life." *Mt Hood Community College Biology 101*, Open Oregon Educational Resources, openoregon.pressbooks.pub/mhccbiology101/chapter/properties-of-life/. Accessed 12 Feb. 2024.

Bazelon, Emily. "The Place of Women on the Court." *The New York Times Magazine*, 7 July 2009.

Beral, Valerie et al. "Breast cancer and abortion: collaborative reanalysis of data from 53 epidemiological studies, including 83,000 women with breast cancer from 16 countries." *Lancet*, vol. 363, no. 9414, U.S. National Library of Medicine, 27 Mar. 2004, https://pubmed.ncbi.nlm.nih.gov/15051280/.

Bergh, Christina, et al. "Obstetric outcome and psychological follow-up of pregnancies after embryo reduction." *Human Reproduction*, vol. 14, no. 8, Oxford Academic, 1 Aug. 1999, https://academic.oup.com/humrep/article/14/8/2170/2913308.

Beriwal, Sridevi, et al. "Multifetal pregnancy reduction and selective termination." *The Obstetrician & Gynaecologist*, vol. 22, no. 4, Wiley, 12 July 2020, https://obgyn.onlinelibrary.wiley.com/doi/10.1111/tog.12690.

Berkowitz, R. S. et al. "Case-control study of risk factors for partial molar pregnancy." *American Journal of Obstetrics and Gynecology*, vol. 173, no. 3 pt. 1, U.S. National Library of Medicine, Sep. 1995, https://pubmed.ncbi.nlm.nih.gov/7573245/.

Biggs, M Antonia et al. "Understanding why women seek abortions in the US." *BMC women's health* vol. 13, no. 29, Springer Nature, 5 Jul. 2013, https://bmcwomenshealth.biomedcentral.com/articles/10.1186/1472-6874-13-29.

"Birth Control Patch." *Mayo Clinic*, Mayo Foundation for Medical Education and Research, 9 Feb. 2023, www.mayoclinic.org/tests-procedures/birth-control-patch/about/pac-20384553.

"Body Functions & Life Process." *Body Functions & Life Process | SEER Training*, NIH, training.seer.cancer.gov/anatomy/body/functions.html. Accessed 12 Feb. 2024.

Bongaarts, John, and Christophe Guilmoto. "How Many More Missing Women? Excess Mortality and Prenatal Sex Selection, 1970–2050." *JSTOR*, vol. 41, no. 2, Population and Development Review, June 2015, www.jstor.org/stable/24639357.

Borenstein, R et al. "Early complications and sequence of pregnancy interruption with hypertonic saline." *International journal of fertility* vol. 25, no. 2, U.S. National Library of Medicine, 1980, https://pubmed.ncbi.nlm.nih.gov/6117532/.

Borgatta, Lynn. "Labor Induction Termination of Pregnancy | GLOWM." *The Alliance for Global Women's Medicine*, FIGO International Federation of Gynecology and Obstetrics, Dec. 2011, www.glowm.com/section-view/heading/Labor Induction Termination of Pregnancy/item/443.

Bransfield, Brian. "Ethical and Religious Directives for Catholic Health Care ..." *USCCB*, United States Conference of Catholic Bishops, June 2018, www.usccb.org/resources/ethical-religious-directives-catholic-health-service-sixth-edition-2016-06_0.pdf.

Brenner, P H et al. "Therapeutic abortions. A review of 567 cases." *California medicine* vol. 115, no. 1, U.S. National Library of Medicine, July 1971, https://pmc.ncbi.nlm.nih.gov/articles/PMC1517925/.

"Brief for the American Association of Pro-Life Obstetricians and Gynecologists as Amicus Curiae," Dobbs v. Jackson, No. 19-1392, (2021) https://www.supremecourt.gov/DocketPDF/19/19-1392/185350/20210729163532595_No.%2019-1392%20-%20American%20Association%20of%20Pro-Life%20Obstetricians%20and%20Gynecologists%20-%20Amicus%20Brief%20in%20Support%20of%20Petitioner%20-%207-29-21.pdf.

Brind, Joel et al. "Induced abortion as an independent risk factor for breast cancer: a systematic review and meta-analysis of studies on south asian women." *Issues of Law and Medicine*, vol. 33, no

1, U.S. National Library of Medicine, Spring 2018, https://pub-
med.ncbi.nlm.nih.gov/30831018/.

Bruder, Carl E G, et al. "Phenotypically Concordant and Discordant
Monozygotic Twins Display Different DNA Copy-Number-
Variation Profiles." *American Journal of Human Genetics*, vol.
82, no. 3, U.S. National Library of Medicine, 29 Feb. 2008,
www.ncbi.nlm.nih.gov/pmc/articles/PMC2427204/.

Burton Brown, Kristi. Fetal Disposition: The Abuses and the Law.
Charlotte Lozier Institute's American Report Series, Issue 13, Dec.
2016. https://lozierinstitute.org/wp-content/uploads/2016/12/
ARS_FetalDisposition_final.pdf

"Care Net Study of American Men whose Partner has had an Abortion." *Lifeway Research*. Carenet, 2021, https://care-net.org/
mens-survey/.

Casey, Frances E. "Abortion - Women's Health Issues." *Merck Manuals Consumer Version*, Merck Manuals, Aug. 2023,
www.merckmanuals.com/home/women-s-health-issues/fam-
ily-planning/abortion.

Casey, Frances E. "Induced Abortion." *Merck Manuals Professional
Edition*, Merck Manuals, Sept. 2023, www.merckmanu-
als.com/professional/gynecology-and-obstetrics/family-plan-
ning/induced-abortion.

Cassidy, S.B. et al. "Five month extrauterine survival in a female triploid (69,XXX) child." *Annals of Human Genetics*, vol. 20, no. 4,
U.S. Library of Medicine, Dec. 1977, https://pub-
med.ncbi.nlm.nih.gov/305757/.

Catholic Church. *Catechism of the Catholic Church: Revised in Accordance with the Official Latin Text Promulgated by Pope John Paul II.* United States Catholic Conference, 2000.

"Catholic Physician Search." *Catholic Medical Association*, Catholic Medical Association, 28 Aug. 2023, www.cathmed.org/physician-directory/.

Cavaliere, Alessandro et al. "Management of molar pregnancy." *Journal of Prenatal Medicine* vol. 3, no. 1, U.S. Library of Medicine, Jan.-Mar. 2009, https://pmc.ncbi.nlm.nih.gov/articles/PMC3279094/.

"Cell Division - Health Video: Medlineplus Medical Encyclopedia." Edited by David Dugdale, *MedlinePlus*, U.S. National Library of Medicine, 7 Sept. 2021, medlineplus.gov/ency/anatomyvideos/000025.htm.

"Cellular Respiration." *HyperPhysics*, Georgia State University, Department of Physics and Astronomy, hyperphysics.phyastr.gsu.edu/hbase/Biology/celres.html. Accessed 27 Feb. 2023.

"CFR - Code of Federal Regulations Title 21." *Accessdata.Fda.Gov*, U.S. Food and Drug Administration, 17 Jan. 2023, www.accessdata.fda.gov/scripts/cdrh/cfdocs/cfcfr/cfr-search.cfm?fr=884.5070.

Charlton, Brittany et al. "Sexual Orientation Differences in Pregnancy and Abortion Across the Lifecourse." *Women's Health Issues*, vol. 30, no. 2, Elsevier Inc., 4 Dec. 2019, www.whijournal.com/article/S1049-3867(19)30483-9/fulltext.

Cheang, Chong-U et al. "A comparison of the outcomes between twin and reduced twin pregnancies produced through assisted reproduction." *Fertility and sterility* vol. 88, no. 1, U.S. National

Library of Medicine, July 2007, https://pubmed.ncbi.nlm.
nih.gov/17270181/.

Chin, Jennifer et al. "Ketamine Compared with Fentanyl for Surgical
Abortion: A Randomized Controlled Trial." *Obstetrics and Gy-
necology*, vol. 140, no. 3, U.S. National Library of Medicine, 1
Sept. 2022, https://pubmed.ncbi.nlm.nih.gov/35926204/.

Chiu, Doris, et al. "As Many as 16% of People Having Abortions Do
Not Identify as Heterosexual Women." *Guttmacher Institute*,
Guttmacher Institute, 7 Dec. 2023, www.guttmacher.org/2023/
06/many-16-people-having-abortions-do-not-identify-hetero-
sexual-women.

"Clinical Practice Handbook for Safe Abortion." *National Library of
Medicine*, World Health Organization, 2014, www.ncbi.nlm.
nih.gov/books/NBK190095/pdf/Bookshelf_NBK190095.pdf.

Clow, W. M., and A. C. Crompton. "The wounded uterus: Pregnancy after hysterotomy." *BMJ*, vol. 1, no. 321, BMJ Publishing
Group Ltd, 10 Feb. 1973, https://www.bmj.com/content/1/
5849/3211.

Coleman, Priscilla K et al. "Late-term elective abortion and suscep-
tibility to posttraumatic stress symptoms." *Journal of Pregnancy*,
U.S. National Library of Medicine, 1 AUg. 2010,
https://pmc.ncbi.nlm.nih.gov/articles/PMC3066627/.

Cómitre-Mariano, Blanca et al. "Feto-maternal microchimerism:
Memories from pregnancy." *iScience* vol. 25, no. 1, U.S. National
Library of Medicine, 29 Dec. 2021, https://pubmed.ncbi.nlm.
nih.gov/35072002/.

"Conception: Fertilization, Process & When It Happens." *Cleveland Clinic*, 6 Sept. 2022, my.clevelandclinic.org/health/articles/11585-conception.

Condic, Maureen L, and Donna Harrison. "Treatment of an Ectopic Pregnancy: An Ethical Reanalysis." *The Linacre Quarterly*, vol. 85, no. 3, U.S. National Library of Medicine, 18 June 2018, www.ncbi.nlm.nih.gov/pmc/articles/PMC6161225/.

Cooper, Danielle B, and Gary W Menefee. "Dilation and Curettage." *National Library of Medicine: National Center for Biotechnology Information*, National Library of Medicine, 9 Mar. 2022, www.ncbi.nlm.nih.gov/books/NBK568791/.

Creinin, M D. "Medical abortion with methotrexate 75 mg intramuscularly and vaginal misoprostol." *Contraception* vol. 56, no. 6, Elsevier, Dec. 1997, https://www.sciencedirect.com/science/article/abs/pii/S001078249700173X

Cue, Lauren, et al. "Hydatidiform Mole." *StatPearls [Internet]*, U.S. National Library of Medicine, 11 Dec. 2024, https://www.ncbi.nlm.nih.gov/books/NBK459155/.

Daling, J.R. et al. "Risk of breast cancer among young women: relationship to induced abortion." *Journal of the National Cancer Institute*, vol. 86, no. 21, U.S. National Library of Medicine, 2 Nov. 1994, https://pubmed.ncbi.nlm.nih.gov/7932822/.

Daneshjou, Khadije, et al. "Congenital Insensitivity to Pain and Anhydrosis (CIPA) Syndrome; a Report of 4 Cases." *Iranian Journal of Pediatrics*, vol. 22, no. 3, U.S. National Library of Medicine, Sept. 2012, pubmed.ncbi.nlm.nih.gov/23400697/.

Dawe, Gavin S, et al. "Cell Migration from Baby to Mother." *Cell Adhesion & Migration*, vol. 1, no. 1, U.S. National Library of

Medicine, 2007, www.ncbi.nlm.nih.gov/pmc/articles/PMC2633
676/#:~:text=Fetal%20cells%20mi-
grate%20into%20the,bone%20mar-
row%2C%20skin%20and%20liver.

De Franciscis, Pasquale et al. "A partial molar pregnancy associated
with a fetus with intrauterine growth restriction delivered at 31
weeks: a case report." *Journal of Medical Case Reports,* vol. 13,
no. 1, U.S. National Library of Medicine, 4 July 2019,
https://pubmed.ncbi.nlm.nih.gov/31269962/.

Delgado, George, et al. "A Case Series Detailing the Successful Reversal of the Effects of Mifepristone Using Progesterone." *Issues
in Law & Medicine*, vol. 33, no. 1, U.S. National Library of Medicine, Spring 2018. https://pubmed.ncbi.nlm.nih.gov/3083
1017/.

Dellapenna, Joseph. *Dispelling the Myths of Abortion History*. Carolina Academic Press, 2006.

Derbyshire, Stuart W G. "Can fetuses feel pain?" *BMJ (Clinical research ed.)* vol. 332, BMJ Publishing Group Ltd, 6 Mar. 2006,
https://www.bmj.com/content/332/7546/909#:~:text=This%20
paper%20discusses%20whether%20there,a%20fetus%20can
%20feel%20pain.

Derbyshire, Stuart WG, and John C Bockmann. "Reconsidering fetal
pain." *Journal of Medical Ethics*, vol. 46, no. 1, BMJ Publishing
Group Ltd & Institute of Medical Ethics, 14 Jan. 2020,
https://jme.bmj.com/content/46/1/3.

"Descriptions of Methods of Pregnancy Termination." *Minnesota
Department of Health*, Minnesota.gov, www.health.state.mn.us/

data/mchs/pubs/abrpt/docs/terminationmethoddescrip-tions.pdf. Accessed 23 Feb. 2023.

Dikke, Galina, and Vladimir Ostromenskiy. 'Interruption of Pregnancy in Women with the Uterine Scar: Potential Risks'. Induced Abortion and Spontaneous Early Pregnancy Loss - Focus on Management, IntechOpen, 22 Apr. 2020. Crossref, doi:10.5772/intechopen.86282.

"Dilation and Evacuation (D&E)." *SOM - State of Michigan*, Michigan Department of Health and Human Services, www.michigan.gov/mdhhs/adult-child-serv/informedconsent/michigans-informed-consent-for-abortion-law/procedures/dilation-and-evacuation-de. Accessed 23 May 2023.

Dobbs v. Jackson Women's Health Organization. U.S. Supreme Court. 2022. Rpt. in "2021 Term Opinions of the Court." *Supreme Court of the United States*, https://www.supremecourt.gov/opinions/21pdf/597us1r58_gebh.pdf

Dobson, Roger. "High cost of abortion is associated with lower pregnancy rate, US study finds." *BMJ*, vol. 336, BMJ PUblishing Group Ltd, 21 Feb. 2008, https://www.bmj.com/content/336/7641/413.5.

Doe v. Bolton. U.S. Supreme Court. 1973. Rpt. in "United States Reports." *Library of Congress*, https://tile.loc.gov/storage-services/service/ll/usrep/usrep410/usrep410179/usrep410179.pdf

Dolapcioglu, Kenan et al. "Twin pregnancy with a complete hydatidiform mole and co-existent live fetus: two case reports and review of the literature." *Archives of Gynecology and Obstetrics*, vol. 279, no. 3, U.S. National Library of Medicine, 5 Aug. 2008, https://pubmed.ncbi.nlm.nih.gov/18679699/.

Donovan, Mary. "Embryology, Weeks 6-8." *StatPearls [Internet].,* U.S. National Library of Medicine, 10 Oct. 2022, www.ncbi.nlm.nih.gov/books/NBK563181/#:~:text=Key%20e mbryologic%20mechanisms%20during%20weeks,disrupted%20by%20failed%20embryologic%20events.

Donovan, Mary F. "Embryology, Yolk Sac." *StatPearls [Internet].,* U.S. National Library of Medicine, 6 Mar. 2023, www.ncbi.nlm.nih.gov/books/NBK555965/.

Dumitrascu, Mihai Christian et al. "The Chemical Pregnancy." *Revista de Chime,* vol. 70, no. 11, Research Gate, Dec. 2019, https://www.researchgate.net/publication/339282838_The_Chemical_Pregnancy.

Duncan, Clara. "Utility of a Routine Ultrasound for Detection of Ectopic Pregnancies among Women Requesting Abortion: A Retrospective Review." *BMJ Sexual & Reproductive Health,* vol. 48, no. 1, U.S. National Library of Medicine, 29 Dec. 2020, pubmed.ncbi.nlm.nih.gov/33376099/.

"Ectopic Pregnancy." *Cleveland Clinic,* Cleveland Clinic, my.clevelandclinic.org/health/diseases/9687-ectopic-pregnancy. Accessed 12 Feb. 2024.

"Ectopic Pregnancy." *Mayo Clinic,* Mayo Foundation for Medical Education and Research, www.mayoclinic.org/diseases-conditions/ectopic-pregnancy/diagnosis-treatment/drc-20372093. Accessed 12 Feb. 2024.

Edelman, A, and N. Kapp. *Dilatation & Evacuation (D&E) Reference Guide: Induced Abortion and Postabortion Care at or after 13 Weeks Gestation ('Second Trimester').* Ipas, 2018.

Edström, K. "Techniques of induced abortion, their health implications and service aspects: a review of the literature." *Bulletin of the World Health Organization* vol. 57, no. 3, U.S National Library of Medicine, 1979, https://pmc.ncbi.nlm.nih.gov/articles/PMC2395811/.

Edstrom, Karin, and Ingegerd Odar-Cederlof. "Therapeutic Abortion by Means of Intrauterine Instillation of Hypertonic Saline." *International Journal of Gynecology & Obstetrics*, vol. 12, no. 2, Wiley, Mar. 1974, https://obgyn.onlinelibrary.wiley.com/doi/abs/10.1002/j.1879-3479.1974.tb00917.x.

Embrey, M.P., and Keith Hillier. "Therapeutic Abortion by Intrauterine Instillation of Prostaglandins." *British Medical Journal*, vol. 1, no. 5749, U.S. National Library of Medicine, 13 Mar. 1971, https://www.ncbi.nlm.nih.gov/pmc/articles/PMC1795245/.

Endler, M, et al. "The Use of Telemedicine Services for Medical Abortion." *Cochrane Database of Systematic Reviews*, 2020.

Evans, Mark I., et al. "Fetal Reduction: 25 Years' Experience." *Karger Publishers*, vol. 35, no. 2, S. Karger AG, 13 Feb. 2014, karger.com/fdt/article/35/2/69/136876/Fetal-Reduction-25-Years-Experience.

"Fact Sheet: LGBTQ+ People & Roe v. Wade." *Amazonaws*, Human Rights Campaign Foundation, hrc-prod-requests.s3-us-west-2.amazonaws.com/FACT-SHEET_-LGBTQ-PEOPLE-ROE-V-WADE.pdf. Accessed 13 Feb. 2024.

Ferguson, Sian. "What Is a Hypertonic-Saline-Induced Abortion?" *Healthline*, Healthline Media, 4 Aug. 2023, www.healthline.com/health/saline-abortion.

"Fetal Awareness: Review of Research and Recommendations for Practice." *Royal College of Obstetricians and Gynaecologists*, Mar. 2010, www.rcog.org.uk/media/xujjh2hj/rcogfetalaware-nesswpr0610.pdf.

"Fetal Development: What Happens during the 3rd Trimester?" *Mayo Clinic*, Mayo Foundation for Medical Education and Research, 3 June 2022, www.mayoclinic.org/healthy-lifestyle/pregnancy-week-by-week/in-depth/fetal-development/art-20045997.

"Find a Pro-Life OBGYN." *AAPLOG*, AAPLOG, aaplog.org/find-a-pro-life-obgyn-search/. Accessed 23 Feb. 2024.

Finer, Lawrence, et al. "Reasons U.S. Women Have Abortions: Quantitative and Qualitative Perspectives." *Guttmacher Institute*, Perspectives on Sexual and Reproductive Health, Sept. 2005, www.guttmacher.org/journals/psrh/2005/reasons-us-women-have-abortions-quantitative-and-qualitative-perspectives.

Gardner, David, and Phil, D. "Development of In Vitro Fertilization Culture Media and the Importance of Blastocyst Transfer." Fertility and Sterility, vol. 110, no. 2, Elsevier Inc., 15 July 2018, https://www.fertstert.org/article/S0015-0282%2818%2930440-0/fulltext.

Genbacev, Olga, et al. "The Role of Chorionic Cytotrophoblasts in the Smooth Chorion Fusion with Parietal Decidua." *Placenta*, vol. 36, no. 7, U.S. National Library of Medicine, 9 May 2015, pubmed.ncbi.nlm.nih.gov/26003500/.

"Gestational Development and Capacity for Pain." *ACOG*, ACOG. www.acog.org/advocacy/facts-are-important/gestational-development-capacity-for-pain. Accessed 21 Feb. 2024

Ghalandarpoor-Attar, Seyedeh Noushin, and Syedeh Mojgan Ghalandarpoor-Attar. "Partial molar pregnancy with a normal live fetus and umbilical cord abnormalities: A novel association with long-term follow-up: A case report." *Clinical Case Reports*, vol. 9, no. 9, U.S. Library of Medicine, 15 Sep. 2021, https://pmc.ncbi.nlm.nih.gov/articles/PMC8443415/.

"Glossary." *NC SCHS: Statistics and Reports: Glossary*, NC Department of Health and Human Services, schs.dph.ncdhhs.gov/data/glossary.htm. Accessed 30 June 2023.

Gouyon, J. B. et al. "[Homogeneous triploid in 2 premature infants (69 XXY)]." *Journal de Genetique Humaine*, vol. 31, no. 4, U.S. National Library of Medicine, Dec. 1983, https://pubmed.ncbi.nlm.nih.gov/6663290/.

Grimes, D. A., and K. F. Schulz. "Morbidity and mortality from second-trimester abortions." *The Journal of reproductive medicine* vol. 30, no. 7, U.S. National Library of Medicine, July 1985, https://pubmed.ncbi.nlm.nih.gov/3897528/.

Guo, Jun et al. "Association between abortion and breast cancer: an updated systematic review and meta-analysis based on prospective studies." *Cancer Causes Control*, vol. 26, no. 6, U.S. National Library of Medicine, 17 Mar. 2015, https://pubmed.ncbi.nlm.nih.gov/25779378/.

Hamed, Khalid M et al. "Overview of Methotrexate Toxicity: A Comprehensive Literature Review." *Cureus* vol. 14, no. 9, U.S.

National Library of Medicine, 23 Sep. 2022, https://pub-med.ncbi.nlm.nih.gov/36312688/.

Hammond, Cassing. "Recent advances in second-Trimester abortion: An evidence-based review." *American Journal of Obstetrics and Gynecology*, vol. 200, no. 4, Elsevier Inc. Apr. 2009, https://www.ajog.org/article/S0002-9378(08)02214-X/fulltext.

Hasegawa, T. et al. "Digynic triploid infant surviving for 46 days." *American Journal of Medical Genetics*, vol. 87, no. 4, U.S. Library of Medicine, 3 Dec. 1999, https://pubmed.ncbi.nlm.nih.gov/10588835/.

Hatcher, Robert, and Deborah Kowal. *Birth Control.* In: Clinical Methods: The History, Physical, and Laboratory Examinations. 3rd edition. Boston: Butterworths; 1990. Chapter 174. Available from: https://www.ncbi.nlm.nih.gov/books/NBK283/

"HB 481/AP: Living Infants Fairness and Equality (LIFE) Act." *Georgia General Assembly*, The Georgia General Assembly, 7 May 2019, www.legis.ga.gov/api/legislation/document/2019 2020/187013.

Healthwise Staff. "Effectiveness Rate of Birth Control Methods." *HealthLink BC British Columbia*, HealthLink BC , 2 Aug. 2022, www.healthlinkbc.ca/pregnancy-parenting/birth-control/effec-tiveness-rate-birth-control-methods.

Healthwise Staff. "How Birth Control Methods Prevent Pregnancy." *HealthLink BC British Columbia*, HealthLink BC, 2 Aug. 2022, www.healthlinkbc.ca/pregnancy-parenting/birth-control/how-birth-control-methods-prevent-pregnancy.

Hegde, Vishwajit S., and Shivaraj Nagalli. "Leucovorin." *StatPearls [Internet].*, U.S. National Library of Medicine, 3 July 2023, www.ncbi.nlm.nih.gov/books/NBK553114/.

Hemida, Reda et al. "Molar Pregnancy with a coexisting living fetus: a case series." *BMC Pregnancy and Childbirth*, vol. 22, no. 1, U.S. Library of Medicine, 3 Sep. 2022, https://pubmed.ncbi.nlm.nih.gov/36057566/.

"Homeostasis." *American Society for Biochemistry and Molecular Biology*, American Society for Biochemistry and Molecular Biology, www.asbmb.org/education/core-concept-teaching-strategies/foundational-concepts/homeostasis. Accessed 12 Feb. 2024.

Horn, Trent. "Questions about Fratelli Tutti, the Souls of Clones, and Abstinence until Marriage." *Catholic Answers*, Catholic Answers, 10 Dec. 2020, www.catholic.com/audio/cot/questions-about-fratelli-tutti-the-souls-of-clones-and-abstinence-until-marriage.

Hospira, Inc. *Full Prescribing Information: Methotrexate Injection*, Pfizer, Inc., 2021.

Howden, Charlotte. "Pregnancy Made Me so Sick That I Begged for an Abortion." *Metro*, Metro.co.uk, 22 July 2020, metro.co.uk/2020/07/22/hyperemesis-gravidarum-abortion-13021924/.

"How Does Pregnancy Happen?: Pregnancy Symptoms & Signs." *Planned Parenthood*, Planned Parenthood Federation of America Inc., www.plannedparenthood.org/learn/pregnancy/how-pregnancy-happens. Accessed 12 Feb. 2024.

"How Does the Abortion Pill Work?: Abortion Pill Function." *Planned Parenthood,* Planned Parenthood Federation of America Inc., https://www.plannedparenthood.org/learn/abortion/the-abortion-pill/how-does-the-abortion-pill-work. Accessed 13 Feb. 2025.

"How to Labour in Water or Have a Water Birth." *NCT (National Childbirth Trust),* 27 Mar. 2020, www.nct.org.uk/labour-birth/different-types-birth/water-birth/how-labour-water-or-have-water-birth.

Hoyert, Donna. "Maternal Mortality Rates in the United States, 2023." *National Center for Health Statistics,* CDC, Feb. 2025, https://www.cdc.gov/nchs/data/hestat/maternal-mortality/2023/Estat-maternal-mortality.pdf.

Hsieh, C.C. et al. "Delivery of a severely anaemic fetus after partial molar pregnancy: clinical and ultrasonographic findings." *Human Reproduction (Oxford, England),* vol. 14, no. 4, U.S. Library of Medicine, April 1999, https://pubmed.ncbi.nlm.nih.gov/10221252/.

Huanxiao, Zhang, et al. "Transvaginal Hysterotomy for Cesarean Scar Pregnancy in 40 Consecutive Cases." *Gynecological Surgery,* vol. 12, no. 1, U.S. National Library of Medicine, 2015, www.ncbi.nlm.nih.gov/pmc/articles/PMC4349961/#:~:text=In%20summary%2C%20this%20transvaginal%20hysterotomy,%2C%20blood%20loss%2C%20and%20cost.

Huchon, Cyrille et al. "Operative hysteroscopy versus vacuum aspiration for incomplete spontaneous abortion (HY-PER): study protocol for a randomized controlled trial." *Trials* vol. 329, no.

14, U.S. National Library of Medicine, 19 Aug. 2015, https://pubmed.ncbi.nlm.nih.gov/37039805/.

Hufbauer, Ellen, et al. *TEACH Early Abortion Training Workbook*. 5th ed., UCSF Bixby Center for Global Reproductive Health , 2016, https://www.teachtraining.org/trainingworkbook/Early-Abortion-Training-Workbook-2016-2.pdf, Accessed 6 June 2023.

"Hysterotomy Abortion." *Wikipedia*, Wikimedia Foundation, 24 Apr. 2023, en.wikipedia.org/wiki/Hysterotomy_abortion.

"Identical Twins Are Not Identical." *Office for Science and Society*, McGill University, 24 Jan. 2021, www.mcgill.ca/oss/article/general-science/identical-twins-are-not-identical.

"If You Are Pregnant: Information on Fetal Development, Abortion and Alternatives." *Minnesota Department of Health*, Minnesota.gov, Dec. 2022, www.lrl.mn.gov/docs/2006/mandated/060480.pdf.

Iliopoulos, Dimitrios et al. "Long survival in a 69,XXX triploid infant in Greece." *Genetics and Molecular Research*, vol. 4, no. 4, U.S. National Library of Medicine, 30 Dec. 2005, https://pubmed.ncbi.nlm.nih.gov/16475122/.

"Induction Abortion." *Kaiser Permanente*, Kaiser Foundation Health Plan, Inc., 2 Aug. 2022, healthy.kaiserpermanente.org/health-wellness/health-encyclopedia/he.induction-abortion.tw2562.

Infante, Sofia. "Margaret Sanger and the KKK: A Racism Scandal." *Human Life International*, Human Life International, 24 Mar. 2021, www.hli.org/resources/margaret-sanger-kkk/.

Ishida, Junji, et al. "Pregnancy-Associated Homeostasis and Dysregulation: Lessons from Genetically Modified Animal Models." *Journal of Biochemistry*, vol. 150, no. 1, U.S. National Library of Medicine, July 2011, pubmed.ncbi.nlm.nih.gov/21613291/#:~:text=Physiological%20alterations%20occur%20in%20many,maintenance%20of%20homeostasis%20in%20pregnancy.

"Iud Birth Control: Info about Mirena & Paragard Iuds." *Planned Parenthood*, Planned Parenthood Federation of America Inc., www.plannedparenthood.org/learn/birth-control/iud. Accessed 12 Feb. 2024.

Jones, Bonnie, and Tracy Weitz. "Legal Barriers to Second-Trimester Abortion Provision and Public Health Consequences." *American Journal of Public Health*, vol. 99, no. 4, U.S. National Library of Medicine, Apr. 2009, www.ncbi.nlm.nih.gov/pmc/articles/PMC2661467/#:~:text=Second%2Dtrimester%20abortion%20providers%20are,extensive%20biased%2Dcounseling%20provisions).

Jones, Kevin. "Notre Dame 'abortion Doula' Talk Was Unworthy of Catholic University, Local Bishop Laments." *Catholic News Agency*, Catholic News Agency, 22 Mar. 2023, www.catholicnewsagency.com/news/253912/notre-dame-abortion-doula-talk-is-squarely-contrary-to-catholic-principles-bishop-rhoades-laments.

Jones, Richard E, and Kristin H Lopez. *Human Reproductive Biology*. Fourth ed., Academic Press, 2014.

Kaczor, Christopher. "'Philosophy and Theology' Reflections on Dignity." *Digital Commons at Loyola Marymount University and*

Loyola Law School, National Catholic Bioethics Quarterly, winter 2015, digitalcommons.lmu.edu/phil_fac/104/#:~:text=The%20meaning%20of%20human%20dignity,the%20dignity%20of%20a%20person.

Kamath, Ramchandra et al. "A study on risk factors of breast cancer among patients attending the tertiary care hospital, in udupi district." *Indian Journal of Community Medicine.* Vol. 38, no. 2, U.S. National Library of Medicine, Apr. 2013, https://pubmed.ncbi.nlm.nih.gov/23878422/.

Kaur, Harmeet. "It's a Stressful Time to Be an Abortion Doula. but Many Say They Aren't Quitting." *CNN,* Cable News Network, 20 July 2022, www.cnn.com/2022/07/15/health/abortion-doulas-roe-v-wade-wellness-cec/index.html.

Kerenyi, Thomas D., et al. "Hypertonic Saline Instillation." *Second-Trimester Abortion,* Springer Dordrecht, https://link.springer.com/chapter/10.1007/978-94-009-8293-2_7.

Kimport, Katrina. "Is third-trimester abortion exceptional? Two pathways to abortion after 24 weeks of pregnancy in the United States." *Perspectives on Sexual and Reproductive Health,* vol. 54, no. 2, 10, U.S. National Library of Medicine, June 2022, https://pubmed.ncbi.nlm.nih.gov/35403366/.

Kolb, Kendra. "Abortion Can Be Medically Necessary." *Live Action,* Live Action - Pro-life Replies, 30 July 2019, prolifereplies.live-action.org/medically-necessary/.

Krugh, Marissa et al. "Misoprostol." *StatPearls [Internet],* U.S. National Library of Medicine, 11 Dec. 2024, https://pubmed.ncbi.nlm.nih.gov/30969695/.

Kulier, Regina, et al. "Surgical Methods for First Trimester Termination of Pregnancy." *Cochrane Database of Systematic Reviews*, 23 Oct. 2001.

Kyono, K. et al. "When Is the Actual Splitting Time of the Embryo to Develop a Monozygotic Dichorionic Diamniotic (DD) Twins Following a Single Embryo Transfer?" *Fertility and Sterility*, vol. 93, no. 4, Elsevier Inc., Sept. 2011, www.fertstert.org/article/S0015-0282(09)00635-9/abstract.

Lafer, C.Z. and R.L. Neu. "A liveborn infant with tertaploidy." *American Journal of Medical Genetics*, vol. 21, no. 2, U.S. National Library of Medicine, Oct. 1988, https://pubmed.ncbi.nlm.nih.gov/3068989/.

Lalitkumar, S., et al. "Mid-trimester induced abortion: A Review." *Human Reproduction Update*, vol. 13, no. 1, U.S. National Library of Medicine, 17 Oct. 2006, https://pubmed.ncbi.nlm.nih.gov/17050523/.

Lawson, Herschel W., et al. "Abortion Surveillance, United States, 1984-1985 ." *CDC Morbidity and Mortality Weekly Report Surveillance Summary:*, Centers for Disease Control, 1 Sept. 1989, www.cdc.gov/mmwr/preview/mmwrhtml/00001467.htm.

Lee, Mike. "Down Syndrome and Social Capital: Assessing the Costs of Selective Abortion." Joint Economic Committee Republicans - US Senate, March 2022.

Lee, Mike. "The Economic Cost of Abortion." Joint Economic Committee Republicans - US Senate, 15 June 2022.

Lenharo, Mariana. "These Drugs Could Restore a Period before Pregnancy Is Confirmed." *Scientific American*, Scientific American, 25 Oct. 2022, www.scientificamerican.com/article/these-drugs-could-restore-a-period-before-pregnancy-is-confirmed/.

"Life Matters: Abortion RLP 2011." *USCCB*, United States Conference of Catholic Bishops, 2011, www.usccb.org/committees/pro-life-activities/life-matters-abortion-rlp-2011.

Liu, Becky, and Asma Khalil. "Selective fetal reduction in multiple pregnancies." *The Continuous Textbook of WOmen's Medicine Sseries - Obstetrics Module*, vol. 4, The Global Library of Women's Medicine, Feb. 2021, https://doi.org/10.3843/glowm.412603.

Lloyd, M.E., et al. "The effects of methotrexate on pregnancy, fertility and lactation." *QJM*, vol. 92, no. 10, Oxford Academic, 1 Oct. 1999, https://doi.org/10.1093/qjmed/92.10.551.

Lohr, Patricia A, et al. "Chapter 10: Dilatation and Evacuation." *Abortion Care*, Cambridge University Press, 2014.

Maraschio, P. et al. "A liveborn 69,XXX triploid. Origin, X chromosome activity and gene dosage." *Annals of Human Genetics*, vol. 27, no. 2, U.S. Library of Medicine, 1984, https://pubmed.ncbi.nlm.nih.gov/6331797/.

"Maternal Deaths." *World Health Organization*, World Health Organization, www.who.int/data/gho/indicator-metadata-registry/imr-details/4622. Accessed 13 Feb. 2024.

"Maternal Mortality." *World Health Organization*, World Health Organization, www.who.int/news-room/fact-sheets/detail/maternal-mortality. Accessed 13 Feb. 2024.

"Maternal Mortality Rates in the United States, 2021." *Centers for Disease Control and Prevention,* Centers for Disease Control and Prevention, 16 Mar. 2023, www.cdc.gov/nchs/data/hestat/maternal-mortality/2021/maternal-mortality-rates-2021.htm.

Mazouni, Chafika, et al. "Termination of pregnancy in patients with previous cesarean section." *Contraception,* vol. 73, no. 3, U.S. National Library of Medicine, Mar. 2006, https://pubmed.ncbi.nlm.nih.gov/16472563/.

McCann, Allison. "What It Costs to Get an Abortion Now." *The New York Times,* 28 Sept. 2022.

"Methods and Medical Risks." *Louisiana Department of Health,* Louisiana Department of Health, 31 Jan. 2023, ldh.la.gov/page/abortion-methods-medical-risks.

"Methods of Abortion." *SCDHEC,* South Carolina Department of Health and Environmental Control., 2019, scdhec.gov/methods-abortion.

MIFEPREX (Mifepristone) Tablets Label - Accessdata.Fda.Gov, Danco Laboratories, LLC , 2016, www.accessdata.fda.gov/drugsatfda_docs/label/2016/020687s020lbl.pdf.

Millward, Adam. ""Nash is pure joy in a tiny package": Most premature baby born 19 weeks early turns one." Guinness World Records, Guiness World Records, 23 July 2025, https://www.guinnessworldrecords.com/news/2025/7/nash-is-pure-joy-in-a-tiny-package-most-premature-baby-born-19-weeks-early-turns-one.

"Molar Pregnancy." *Better Health Channel,* Victoria State Government Department of Health, 25 Jan. 2024, https://www.betterhealth.vic.gov.au/health/healthyliving/molar-pregnancy.

Moore, Keith. "Fetal Growth and Development." *South Dakota Department of Health*, South Dakota Department of Health, 1995, doh.sd.gov/media/bnemplje/fetal.pdf.

Morton, Sarah, and Dara Brodsky. "Fetal Physiology and the Transition to Extrauterine Life." *Clinics in Perinatology*, vol. 43, no. 3, U.S. National Library of Medicine, 11 June 2016, www.ncbi.nlm.nih.gov/pmc/articles/PMC4987541/.

"Motion and Brief Amicus Curiae of Certain Physicians, Professors and Fellows of the American College of Obstetrics and Gynecology in Support of Appellees," Roe v. Wade, No. 1-53, (1971).

Myers, Laura B., et al. "Fetal endoscopic surgery: indications and anaesthetic management." *Best Practice & Research Clinical Anaesthesiology*, vol. 18, no. 2, Elsevier, June 2004, https://www.sciencedirect.com/science/article/abs/pii/S1521689604000023.

"Naprotechnology." *The Result of 30 Years of Research and Education*, NAPRO, naprotechnology.com/. Accessed 23 Feb. 2024.

Natoli, Jaime, et al. "Prenatal Diagnosis of Down Syndrome: A Systematic Review of Termination Rates (1995-2011)." *Prenatal Diagnosis*, John Wiley & Sons, 14 Mar. 2012, obgyn.onlinelibrary.wiley.com/doi/10.1002/pd.2910.

"NCI Dictionary of Cancer Terms." *National Cancer Institute*, NIH, www.cancer.gov/publications/dictionaries/cancer-terms/def/cell. Accessed 12 Feb. 2024.

Newmann, Sara J, et al. "Cervical Preparation for Second Trimester Dilation and Evacuation." *Cochrane Database of Systematic Reviews*, 4 Aug. 2010.

Niemann-Seyde, S.C. et al. "A case of full triploidy (69,XXX) of paternal origin with unusually long survival time." *Clinical Genetics*, vol. 43, no. 2, U.S. Library of Medicine, Feb. 1993, https://pubmed.ncbi.nlm.nih.gov/8448906/.

"NIH Grants Policy Statement: 4.1.14 Human Fetal Tissue Research." *National Institutes of Health*, U.S. Department of Health and Human Services, Dec. 2022, grants.nih.gov/grants/policy/nihgps/HTML5/section_4/4.1.14_human_fetal_tissue_research.htm#Human3.

"Non-Invasive Prenatal Screening Tests May Have False Results." *U.S. Food and Drug Administration*, FDA, 19 Apr. 2022, www.fda.gov/medical-devices/safety-communications/genetic-non-invasive-prenatal-screening-tests-may-have-false-results-fda-safety-communication.

"Overview." Advanced Bioscience Resources, Inc, Center for Medical Progress, Acc. 16 Aug. 2025, https://www.centerformedicalprogress.org/wp-content/uploads/2015/05/OVERVIEW.pdf.

Padawer, Ruth. "The Two-Minus-One Pregnancy." *The New York Times*, The New York Times Company, 10 Aug. 2011.

"Parental Involvement in Minors' Abortions." *Guttmacher Institute*, Guttmacher Institute, 1 Sept. 2023, www.guttmacher.org/state-policy/explore/parental-involvement-minors-abortions.

Parker, Star, et al. "The Effects of Abortion on the Black Community." *Center for Urban Renewal and Education*, June 2015, cure-policy.org/?s=the%2Beffects%2Bof%2Babortion%2Bon%2Bthe%2Bblack%2Bcommunity%2Bjune%2B2015#.

"Partial Molar Pregnancy." *Cleveland Clinic,* Cleveland Clinic, 11 April 2023, https://my.clevelandclinic.org/health/diseases/12332-partial-molar-pregnancy.

Paul, Maureen, et al. "Early molar pregnancy: experience in a large abortion service." *Contraception,* vol. 81, no. 2, Elsevier, February 2010, https://www.sciencedirect.com/science/article/abs/pii/S0010782409003862.

Pawlowski, A. "Born at 21 Weeks, This May Be the Most Premature Surviving Baby." *TODAY.Com,* NBC Universal, 21 Nov. 2018, www.today.com/health/born-21-weeks-she-may-be-most-premature-surviving-baby-t118610.

Pfizer Labs. *Cytotec Misoprostol Tablets,* Pfizer Inc., New York, NY, 2021.

Piya-Anant, Manee, and Siripong Swasdimongkol. "Outcome of Intra-amniotic Hypertonic Saline Instillation for Second-Trimester Abortion." *Siriraj Medical Journal,* vol. 64, no. 1, Siriraj Medical Journal, Jan.-Feb. 2012, https://he02.tci-thaijo.org/index.php/sirirajmedj/article/view/244143.

"Plan B One-Step Emergency Contraceptive." *Target,* Target Brands Inc., www.target.com/p/plan-b-one-step-emergency-contraceptive/-/A-14847439?ref=tgt_adv_xsp&AFID=google&fndsrc=tgtao&DFA=71700000049304432&CPNG=PLA_Health_Priority%2BShopping%7CHealth_Ecomm_Essentials&adgroup=Health_Priority%2BTCINs&LID=700000001170770pgs&LNM=PRODUCT_GROUP&network=g&device=c&location=9016237&targetid=pla-321577217246&gad_source=1&

gclid=CjwKCAiA29auBhBxEiwAnKcSqsl5o-_0tHSVJ5wozvp-
mAotRhP8OgnL7bbRnpK8z80FsTLIsTA5HLRoCnPQQAvD_
BwE&gclsrc=aw.ds. Accessed 21 Feb. 2024.

"Plan C: Abortion Pills by Mail in Every State." *PLAN C: Abortion Pills by Mail in Every State*, National Women's Health Network, 2023, www.plancpills.org/.

"Plan C: Get abortion pills now, just in case." *Instagram.* 2023, https://www.instagram.com/p/CwQ5tUIIJ_z/?hl=en&img_index=2

Planned Parenthood of Southeastern Pennsylvania v. Casey. U.S. Supreme Court. 1992. Rpt. in *Justia,* https://supreme.justia.com/cases/federal/us/505/833/case.pdf

"Planned Parenthood Staff and Partners Admitted Under Oath to Harvest the Still Beating Hearts of Aborted Babies." Live Action, Instagram, 19 March 2025, https://www.instagram.com/live-actionorg/reel/DHYe8lXxp5q/?api=.

"Pregnancy Mortality Surveillance System." *Centers for Disease Control and Prevention*, Centers for Disease Control and Prevention, 23 Mar. 2023, www.cdc.gov/reproductivehealth/maternal-mortality/pregnancy-mortality-surveillance-system.htm.

"Pregnancy Mortality Surveillance System." *Centers for Disease Control and Prevention*, Centers for Disease Control and Prevention, 14 Nov. 2024, https://www.cdc.gov/maternal-mortality/php/pregnancy-mortality-surveil-lance/?CDC_AAref_Val=https://www.cdc.gov/reproductivehealth/maternal-mortality/pregnancy-mortality-surveil-lance-system.htm.

"Preterm Birth." *World Health Organization*, World Health Organization, 10 May 2023, www.who.int/news-room/fact-sheets/detail/preterm-birth.

"Preventing Gender-Biased Sex Selection." *World Health Organization,* World Health Organization, 2011, apps.who.int/iris/bitstream/handle/10665/44577/9789241501460_eng.pdf;jsessionid=65CFEA592B0AD8032CA96540A6846A0F?sequence=1.

Protocol Recommendations For Use of Methotrexate and Misoprostol in Early Abortion, National Abortion Federation, 2005.

"Quality & Safety." *Ambulatory Surgery Center Association (ASCA)*, 2023, www.ascassociation.org/about-ascs/quality-and-safety.

"Questions and Answers on Mifepristone for Medical Termination of Pregnancy Through Ten Weeks Gestation." *U.S. Food and Drug Administration*, U.S. Food and Drug Administration, 4 Jan. 2023, www.fda.gov/drugs/postmarket-drug-safety-information-patients-and-providers/questions-and-answers-mifepristone-medical-termination-pregnancy-through-ten-weeks-gestation.

Rahamni, Maryam and Sara Parviz. "A case report of partial molar pregnancy associated with a normal appearing dizygotic fetus." *Asian Pacific Journal of Reproduction*, vol. 5, no. 2, ScienceDirect, April 2016, https://www.sciencedirect.com/science/article/pii/S2305050016000245.

Raja, Srinivasa N et al. "The revised International Association for the Study of Pain definition of pain: concepts, challenges, and compromises." *Pain* vol. 161, no. 9, U.S. National Library of

Medicine, 1 Sept. 2020, https://pubmed.ncbi.nlm.nih.gov/ 32694387/.

Ranji, Usha, and Karen Diep. "Key Facts on Abortion in the United States." *Women's Health Policy*, KFF, 11 May 2023, www.kff.org/womens-health-policy/issue-brief/key-facts-on-abortion-in-the-united-states/.

Rao, Suman. "In Utero Fuel Homeostasis: Lessons for a Clinician." *Indian Journal of Endocrinology and Metabolism*, vol. 17, no. 1, U.S. National Library of Medicine, 17 Jan. 2013, pubmed.ncbi. nlm.nih.gov/23776854/.

Rattu, Mohammad A et al. "Glucarpidase (voraxaze), a carboxypeptidase enzyme for methotrexate toxicity." *P & T : a peer-reviewed journal for formulary management* vol. 38, no. 12, U.S. National Library of Medicine, Dec. 2013, https://pmc.ncbi.nlm.nih.gov/ articles/PMC3875266/.

Reardon, David. "The Embrace of the Proabortion Turnaway Study: Wishful thinking? Or Willful Deception." *The Linacre Quarterly*, vol. 85, no. 3,U.S. National Library of Medicine, 20 June 2018, https://pmc.ncbi.nlm.nih.gov/articles/PMC6161227/.

Reardon, David et al. "Overlooked Dangers of Mifepristone, the FDA's Reduced REMS, and Self-Managed Abortion Policies: Unwanted Abortions, Unnecessary Abortions, Unsafe Abortions." *Charlotte Lozier Institute,* Charlotte Lozier Institute, 16 Dec. 2021, https://lozierinstitute.org/overlooked-dangers-of-mifepristone-the-fdas-reduced-rems-and-self-managed-abortion-policies-unwanted-abortions-unnecessary-abortions-unsafe-abortions/.

Rico, S. et al. "[Complete triploidy in a liveborn premature (author's transl)]." *An Esp Pediatr.*, vol. 13, no. 1, U.S. National Library of Medicine, Jan. 1980, https://pubmed.ncbi.nlm.nih.gov/7189 392/.

Richter, Ruthann, "Premature Babies' Survival Rate is Climbing, Study Says." *Stanford Medicine Children's Health,* Stanford Medicine Children's Health, 10 Feb. 2022, https://healthier.stan-fordchildrens.org/en/premature-babies-survival-rate-is-climb-ing/?utm_campaign=feed&utm_medium=refer-ral&utm_source=later-linkinbio.

Roberts, H.E. et al. "Unique mosaicism of tetraploidy and trisomy 8: clinical, cytogenetic, and molecular findings in a live-born infant." *American Journal of Medical Genetics*, vol. 62, no. 3, U.S. Library of Medicine, 29 Mar. 1996, https://pubmed.ncbi.nlm. nih.gov/8882781/.

Roe. v. Wade. U.S. Supreme Court. 1973. Rpt. in "United States Reports." *Library of Congress,* https://tile.loc.gov/storage-ser-vices/service/ll/usrep/usrep410/usrep410113/usrep410113.pdf

Roe. v. Wade, White Dissent. U.S. Supreme Court. 1973. Rpt. in "United States Reports." *Library of Congress,* https://www.law. cornell.edu/supremecourt/text/410/179#writing-USSC_CR_ 0410_0179_ZD

Sajadi-Ernazarova, Karima R., and Christopher L. Martinez. "Abortion Complications." *StatPearls [Internet].,* U.S. National Library of Medicine, 16 May 2023, www.ncbi.nlm.nih.gov/ books/NBK430793/.

Sam, Sreya et al. "Trends of Selective Fetal Reduction and Selective Termination in Multiple Pregnancy, in England and Wales: a

Cross-Sectional Study." *Reproductive sciences (Thousand Oaks, Calif.)* vol. 29, no. 3, U.S. National Library of Medicine, March 2022, https://pubmed.ncbi.nlm.nih.gov/34902100/.

Sanger, Margaret. "The Birth Control Review v.5 1921." *HathiTrust,* Cornell University, Oct. 1921, babel.hathitrust.org/cgi/pt?id= coo.31924007352325&seq=193&q1=urgent.

Santiago-Munoz, Patricia. "The Truth about Ectopic Pregnancy Care: Your Pregnancy Matters: UT Southwestern Medical Center." *Your Pregnancy Matters | UT Southwestern Medical Center,* The University of Texas Southwestern Medical Center, 22 Oct. 2019, utswmed.org/medblog/truth-about-ectopic-pregnancy-care/#:~:text=Patricia%20Santi-ago%2DMu%C3%B1oz%2C%20M.D.,prior%20pel-vic%20or%20abdominal%20surgery.

Sargin, Mehmet Akif and Niyazi Tug. "Prenatal screening tests may be a warning for the parietal molar pregnancy? case report." *The Pan African Medical Journal,* vol. 20, no. 323, U.S. National Library of Medicine, 6 Apr. 2015, https://pubmed.ncbi.nlm. nih.gov/26175814/.

"SB-24 Public Health: Public University Student Health Centers: Abortion by Medication Techniques." *California Legislative Information,* State of California, 14 Oct. 2019, leginfo.legisla-ture.ca.gov/faces/billTextClient.xhtml?bill_id=201920200SB24.

Schmidt-Hansen, Mia et al. "Initiation of abortion before there is definitive ultrasound evidence of intrauterine pregnancy: A sys-tematic review with meta-analyses." *Acta obstetricia et gyneco-*

logica Scandinavica vol. 99, no. 4, U.S. National Library of Medicine, 22 Jan. 2020, https://pubmed.ncbi.nlm.nih.gov/31883371/.

Schneider, A Patrick 2nd et al. "The breast cancer epidemic: 10 facts." *Linacre Quarterly*, vol. 81, no. 3, U.S. National Library of Medicine, Aug. 2014, https://pubmed.ncbi.nlm.nih.gov/25249706/.

Schnettler, William. "Termination of Pregnancy." *Maternal Cardiac Care: A Guide to Managing Pregnant Women with Heart Disease*, Elsevier, 2023.

"Secondary Infertility." Cleveland Clinic, Cleveland Clinic, my.clevelandclinic.org/health/diseases/21139-secondary-infertility. Accessed 21 Feb. 2024.

"Second Trimester Labor Induction Abortion." *SOM - State of Michigan*, Michigan Department of Health and Human Services, 2023, www.michigan.gov/mdhhs/adult-child-serv/informed-consent/michigans-informed-consent-for-abortion-law/procedures/second-trimester-labor-induction-abortion.

Sherard, J. et al. "Long Survival in a 69,XXY triploid male." *American Journal of Medical Genetics,* vol. 25, no. 2, U.S. Library of Medicine, Oct. 1986, https://pubmed.ncbi.nlm.nih.gov/3777027/#:~:text=Abstract,with%20this%20condition%20to%20date..

Shulman, L P et al. "Dilation and evacuation for second-trimester genetic pregnancy termination." *Obstetrics and gynecology* vol. 75, no. 6, U.S. National Library of Medicine, June 1990, https://pubmed.ncbi.nlm.nih.gov/2342729/.

Siwatch, Sujata, et al. "Hysterotomy- indications and associated complications: An indian teaching hospital experience." *Nepal*

Journal of Obstetrics and Gynaecology, vol. 7, no. 2, Nepal Journals Online, Sept. 2012, https://www.nepjol.info/index.php/NJOG/article/view/11136.

Sloan, Don, and Paula Hartz. *Choice: A Doctor's Experience with the Abortion Dilemma.* International Publishers, 2002.

Smoley, Brian A., and Christa M. Robinson. "Natural Family Planning." *American Family Physician*, vol. 86, no. 10, American Academy of Family Physicians, 15 Nov. 2012, https://aafp.org/pubs/afp/issues/2012/1115/p924.html#:~:text=The%20five%20principal%20types%20of,amenorrhea4%20(Table%201).

Soleimani Movahed, Maryam, et al. "The economic burden of abortion and its complication treatment cares: A systematic review." *Journal of Family & Reproductive Health*, vol. 14, no. 2, kne publishing, 7 Oct. 2020, https://publish.kne-publishing.com/index.php/JFRH/article/view/4354.

"Stages of Fetal Development - First Trimester: LA Dept. of Health." *Stages of Fetal Development - First Trimester | La Dept. of Health*, Louisiana Department of Health, ldh.la.gov/page/stages-of-fetal-development-first-trimester. Accessed 12 Feb. 2024.

"Stages of Fetal Development - Third Trimester: LA Dept. of Health." *Stages of Fetal Development - Third Trimester | La Dept. of Health*, Louisiana Department of Health, ldh.la.gov/page/stages-of-fetal-development-third-trimester. Accessed 12 Feb. 2024.

Stenberg v. Carhart. United States Supreme Court. 2000. Rpt. in "United States Reports." *Library of Congress, https://tile.loc.gov/storage-services/service/ll/usrep/usrep530/usrep530914/usrep530914.Pdf.*

Stubblefield, Phillip G et al. "Methods for induced abortion." *Obstetrics and gynecology* vol. 104, no.1, U.S. National Library of Medicine, July 2004, https://pubmed.ncbi.nlm.nih.gov/15229018/.

Studnicki, James, et al. "Improving Maternal Mortality Comprehensive Reporting for All Pregnancy Outcomes." *SCIRP*, vol. 7, no. 8, Scientific Research Publishing, 23 Aug. 2017, www.scirp.org/journal/paperinformation?paperid=78764.

Studnicki, James, et al. "Improving the Metrics and Data Reporting for Maternal Mortality: A Challenge to Public Health Surveillance and Effective Prevention." *Online Journal of Public Health Informatics*, vol. 11, no. 2, U.S. National Library of Medicine, 19 Sept. 2019, ncbi.nlm.nih.gov/pmc/articles/PMC6788905/.

Studnicki, James. "Late-Term Abortion and Medical Necessity: A Failure of Science." *Health Services Research and Managerial Epidemiology*, vol. 6, U.S. National Library of Medicine, 9 April 2019, https://pmc.ncbi.nlm.nih.gov/articles/PMC6457018/.

Sun, Yuelian, et al. "Induced Abortion and Risk of Subsequent Miscarriage." *International Joiurnal of Epidemiology*, vol. 32, no. 3, Oxford University Press, 1 June 2003, academic.oup.com/ije/article/32/3/449/637113.

Surendran, Smitha et al. "Partial Molar Pregnancy with a Normal Fetus with Complete Placenta Previa." *The Journal of Obstetrics and Gynecology of India*, vol. 68, no. 3, U.S. Library of Medicine, 8 June 2017, https://pmc.ncbi.nlm.nih.gov/articles/PMC5972090/.

Tan, Cheryl, and Adam Lewandowski. "The Transitional Heart: From Early Embryonic and Fetal Development to Neonatal Life." *Fetal Diagnosis and Therapy*, vol. 47, no. 5, S. Karger AG,

18 Sept. 2019, karger.com/fdt/article/47/5/373/136983/The-Transitional-Heart-From-Early-Embryonic-and.

Tanne, Janice Hopkins. "Abortion: Indiana Becomes First US State to Enact an Almost Total Ban." *The BMJ*, British Medical Journal Publishing Group, 10 Aug. 2022, www.bmj.com/content/378/bmj.o1998.

"Targeted Regulation of Abortion Providers." *Guttmacher Institute*, Guttmacher Institute, 31 Aug. 2023, www.guttmacher.org/state-policy/explore/targeted-regulation-abortion-providers.

"The Different Types of Abortions - Abortion Methods." *Eastside Gynecology*, Eastside Gynecology, 26 Sept. 2018, eastsidegyne-cology.com/blog/different-types-of-abortion/.

"The Voyage of Life Week 7." *Lozier Institute*, Charlotte Lozier Institute, 22 Aug. 2023, lozierinstitute.org/fetal-development/week-7/.

"The Voyage of Life: Weeks 21 & 22." *Lozier Institute*, Charlotte Lozier Institute, 22 Aug. 2023, lozierinstitute.org/fetal-development/weeks-21-and-22/.

Thill, Bridget. "Fetal Pain in the First Trimester." *The Linacre Quarterly*, vol. 89, no. 1, Sage Journals / Catholic Medical Association, 6 Dec. 2021, https://journals.sagepub.com/doi/10.1177/00243639211059245.

"Think Before You Drink: Source Guide." *Students for Life,* This is Chemical Abortion, Accessed 30 Apr. 2025, https://thisischem-icalabortion.com/sources/.

Title IX, Education Amendments of 1972, 20 U.S.C. §§ 1681-1688.

United States, Congress. Public Law 108-105. Partial-Birth Abortion Ban Act. congress.gov, 2003. Congress.gov. https://www.congress.gov/108/plaws/publ105/PLAW-108publ105.pdf

Vergaro, Paula, and Roger Sturmey. "Embryo Metabolism and What Does the Embryo Need? (Chapter 4) - Manual of Embryo Culture in Human Assisted Reproduction." *Cambridge Core,* Cambridge University Press, 15 Apr. 2021, www.cambridge.org/core/books/abs/manual-of-embryo-culture-in-human-assisted-reproduction/embryo-metabolism-and-what-does-the-embryo-need/7A2F713FA4F68591CB2162A85E7601E5.

Wang, Chao et al. "The clinical outcomes of selective and spontaneous fetal reduction of twins to a singleton pregnancy in the first trimester: a retrospective study of 10 years." *Reproductive Biology and Endocrinology* vol. 20, no. 71, Springer Nature, 22 Apr. 2022, https://rbej.biomedcentral.com/articles/10.1186/s12958-022-00935-0.

Wang, James, et al. "Hysterectomy." *Women and Health (Second Edition),* Academic Press, 7 Dec. 2012, www.sciencedirect.com/science/article/abs/pii/B9780123849786000273.

"Week 34." *NHS Choices,* NHS, https://www.nhs.uk/start-for-life/pregnancy/week-by-week-guide-to-pregnancy/3rd-trimester/week-34/. Accessed 21 Feb. 2024.

"Week 36." *NHS Choices,* NHS, www.nhs.uk/start-for-life/pregnancy/week-by-week-guide-to-pregnancy/3rd-trimester/week-36/. Accessed 12 Feb. 2024.

"Week 38." *NHS Choices*, NHS, www.nhs.uk/start-for-life/preg-nancy/week-by-week-guide-to-pregnancy/3rd-trimester/week-38/. Accessed 12 Feb. 2024.

"Week 40." *NHS Choices*, NHS, www.nhs.uk/start-for-life/preg-nancy/week-by-week-guide-to-pregnancy/3rd-trimester/week-40/. Accessed 12 Feb. 2024.

Wehrman, Jessica. "State Abortion Bans Bar Exceptions for Suicide, Mental Health." *Roll Call*, CQ Roll Call, 7 Dec. 2023, roll-call.com/2023/12/07/state-abortion-bans-bar-exceptions-for-suicide-mental-health/.

"What Are Examples and Causes of Maternal Morbidity and Mor-tality?" *Eunice Kennedy Shriver National Institute of Child Health and Human Development*, U.S. Department of Health and Human Services, 14 May 2020, www.nichd.nih.gov/health/topics/maternal-morbidity-mortality/conditioninfo/causes.

"What happens when 30 million Chinese men can't find wives? Da-ting coaches step in with special camps" The Economic Times, Bennett, Coleman & Co. Ltd, 18 July 2025, https://econom-ictimes.indiatimes.com/news/new-updates/what-happens-when-30-million-chinese-men-cant-find-wives-dating-coaches-step-in-with-special-camps/arti-cleshow/122768952.cms?from=mdr.

"Where Do Americans Stand on Abortion?" *SBA Pro-Life America*, SBA Pro-Life America, 19 Dec. 2023, sbaprolife.org/polling.

Witwer, Elizabeth, et al. "Abortion service delivery in clinics by state policy climate in 2017." *Contraception: X*, vol. 2, Elsevier, 16 Oct. 2020, https://doi.org/10.1016/j.conx.2020.100043.

"Womans-Right-to-Know.Pdf - Texas Health and Human Services." *Texas Health and Human Services*, Texas Department of State Health Services, Dec. 2016, www.hhs.texas.gov/sites/default/files/documents/services/health/women-children/womans-right-to-know.pdf.

"Yolk Sac in Early Pregnancy: Meaning & Function." *Cleveland Clinic*, Cleveland Clinic, my.clevelandclinic.org/health/body/22341-yolk-sac. Accessed 12 Feb. 2024.

"You and Your Baby at 7 Weeks Pregnant." *NHS Choices*, NHS, 15 Mar. 2023, www.nhs.uk/pregnancy/week-by-week/1-to-12/7-weeks/.

Yuan, Xuelian et al. "Induced Abortion, Birth Control Methods, and Breast Cancer Risk: A Case-Control Study in China." *Journal of Epidemiology*, vol. 29, no. 5, U.S. National Library of Medicine, 5 May 2019, https://pmc.ncbi.nlm.nih.gov/articles/PMC6445797

J.M.J. A.M.D.G.

www.ingramcontent.com/pod-product-compliance
Lightning Source LLC
Chambersburg PA
CBHW052034090426
42739CB00010B/1911